Monograph Series in

UNDERWATER ACOUSTICS

Sponsored by Office of Naval Research

For further volumes:
http://www.springer.com/series/5535

Monograph Series in

UNDERWATER ACOUSTICS

In 2001 the Office of Naval Research began a program to publish a selected monograph series in the field of underwater acoustics. The series publishes in-depth reviews and analysis on the state of understanding of the physics of sound in selected areas of research.

The monograph series covers the most important topics in underwater sound in an in-depth, critical, scholarly, and comprehensive fashion. It is intended to provide researchers with an understanding of the physics of underwater sound, based on both theory and experiment. The discussion of important unanswered questions is encouraged. Topics of particular importance in the field are low-frequency sound in the deep ocean, shallow-water acoustics, and ambient noise.

Sponsored by:

Office of Naval Research
One Liberty Center
875 North Randolph Street, Suite 1425
Arlington, VA 22203-1995
E-mail: onrpao@onr.navy.mil

Series Editors:

Jeffrey A. Simmen, Ph.D.
Applied Physics Laboratory
University of Washington
1013 NE 40th Street
Seattle, WA 98105-6698
E-mail: simmen@apl.washington.edu

Homer P. Bucker, Ph.D.
Space and Naval Warfare Systems Center, retired
808 Moana Drive
San Diego, CA 92106
E-mail: bucker@nosc.mil

Ira Dyer, Ph.D.
Weber Shaughness Professor of Ocean Engineering, Emeritus
Massachusetts Institute of Technology
Cambridge, MA 02139

Finn B. Jensen, Ph.D.
Emeritus Scientist
NATO Undersea Research Centre
19126 La Spezia, Italy
E-mail: jensen@nurc.nato.int

Ellen S. Livingston, Ph.D.
Associate Director
Ocean and Undersea Science
U.S. Office of Naval Research Global
86 Blenheim Crescent
West Ruislip, Middlesex HA4 7HB
United Kingdom
E-mail: ellen.livingston@onrg.navy.mil

William M. Carey
Richard B. Evans

Ocean Ambient Noise

Measurement and Theory

Springer

William M. Carey
Department of Mechanical Engineering
Boston University
Old Lyme, CT 05371, USA
wcarey@bu.edu

Richard B. Evans
North Stonington
CT 06359-1038, USA

ISBN 978-1-4419-7831-8 e-ISBN 978-1-4419-7832-5
DOI 10.1007/978-1-4419-7832-5
Springer New York Dordrecht Heidelberg London

Printed on acid-free paper

Springer is part of Springer Science+Business Media (www.springer.com)

Chapter 1
Introduction

Ambient noise became an active area of research during World War II because of the availability of calibrated instruments and the necessity to understand the ambient noise levels in coastal waters. This wartime research was summarized by Knudsen et al. (1948), and later by Urick and Pryce (1954). After the war, during the 1950s, published ambient noise research waned until the classic paper of Wenz (1962). This work initiated a renaissance of ambient noise as one of the most interesting areas of oceanic acoustic research. The classic paper of Wenz (1962) was notable as it supplied a graphical or schematic spectrum, omnidirectional noise levels versus frequency. This schematic identified sources of ambient noise and resultant omnidirectional levels in frequency bands parameterized by the Beaufort wind force. This metric includes the 10-m wind speed as well as the appearance of the sea. The schematic also identified regions dominated by shipping and regions dominated by rain noise.

Most ambient noise research between 1950 and 1980 was classified; recently, several key experimental papers from this period were published in a special archival issue of the *Journal of Oceanic Engineering*, Carey (2005), and represent important benchmarks of observed noise levels and directionalities. Walkinshaw (2005) presented 4 years of noise measurements in the Norwegian Sea. This work was unique because sensors and recording instrumentation in the period from 1957 through 1961 were rather primitive compared with current technology and the difficulty in performing these measurements cannot be understated. Nichols (2005) obtained noise results between 1951 and 1974 using new specially built barium titanate hydrophones to perform the measurements that stressed transient noise sources (biological, machines, and offshore drilling) compared with the background of wind-driven and shipping noise. He discussed in detail the mysterious 20-Hz/20-cycle sounds correctly attributed to cetaceans. According to Urick (1984), the observation of these sound sequences was so mysterious that their occurrence was highly classified until a 1963 conference on marine bioacoustics and a paper by Walker (1963). Reports and phonographic recordings became available on bioacoustics and marine mammal sounds [Tavolga (1964, 1965), Hills (1968); a knowledge base of these recorded sounds is available from the Historic Naval Ships Association, Smithfield, VA, USA, http://www.hnsa.org].

W.M. Carey, R.B. Evans, *Ocean Ambient Noise*, Underwater Acoustics,
DOI 10.1007/978-1-4419-7832-5_1, © William M. Carey and Richard B. Evans 2011

The ambient noise problem was a primary focus of the scientists assembled by the Office of Naval Research under a special project called LRAPP (Long Range Acoustic Propagation Project). This aim of this project was first to develop a quantitative understanding of propagation and noise and second to develop predictive techniques for calculating ambient noise levels in worldwide ocean areas. LRAPP was under the direction of Bracket Hersey and Roy Gaul. Many ambient noise discoveries were either directly or indirectly a consequence of this program. In 1974, Bracket Hersey held an "International Workshop on Low Frequency Propagation and Noise" at the Woods Hole Oceanographic Institution and published three volumes of the proceedings. The first two volumes were available to the general public, the third was not. Selected papers from the third volume were reviewed and included in Carey (2005). Vertical directionality of the noise field was considered important and measurements found in Garabed (2005) are of archival interest because of their scope and relevance to the continuing interest in noise levels and vertical directive effects. Garabed's work was in a band between 200 and 380 Hz and most of the LRAPP work was at less than 1 kHz. Several additional vertical noise measurements were made public [see Urick (1984), Carey and Wagstaff (1986)].

By 1976, multiple summaries, bibliographies, and a vast amount of literature (about 1,500 references) had been published on the measurements, theory, and computational methods; and by the 1980s, ambient noise was the second largest area of underwater acoustics. In 1984, Urick (1984) summarized the main features of ambient noise from the unclassified literature but stated that a vast amount of classified literature existed. In this review, Urick commented on the prolific nature of literature concerning theories of sound generation at the sea surface and the measurements of the temporal and spatial spectral characteristics. The idealized spectra suggested by Urick were in agreement with the schematic proposal by Wenz (1962). The spectral characteristics for frequencies greater than 500 Hz were also consistent with the observations of Knudsen (1948) and Wenz's "rule of fives": "In the frequency band between 500 Hz to 5 kHz the ambient sea-noise spectrum levels decrease 5 dB per octave with increasing frequency, and increase 5 dB with each doubling of wind speed from 2.5 to 40 knots; the spectrum level at 1 kHz in deep water is equal to 25 dB (5 $\times$ 5) re 0.0002 dyn/cm^2 when the wind speed is 5 knots, and is 5 dB higher in shallow water." This "rule of fives" can be expressed as follows:

$$NL(f, U) = 25 - 10 \cdot log[f^{5/3}] + 10 \cdot log[(U/5)^{5/3}]$$

or

$$NL(f, U) = 25 - (5/3) \cdot 10 \cdot [log[f] - log[U/5]]$$

where f is frequency (kHz) and U is wind speed (knots). However, Wenz observed in the 10–500-Hz band the measured noise levels were often variable and dominated by shipping noise. The shape of the spectrum was also found to vary from a positive slope to a steep negative slope.

Kerman (1984) showed that "the amalgamated observations of the ambient noise reveal a similarity structure, both in the acoustical spectrum and wind dependency."

For frequencies greater than the local maximum in the 300–500-Hz range, Kerman found that the normalized measured spectral characteristic was proportional to f^{-2} (6 dB/octave). Furthermore, he showed that the noise intensity was proportional to the cube of the friction velocity (u_*^3) prior to a critical friction velocity (u_{*c}), which is determined by the minimum phase velocity of the gravity–capillary waves. Wave breaking was associated with this critical condition, and for $u^* > u_{*c}$ the noise intensity was found to increase with $u_*^{1.5}$. These observations were found to be consistent with a large number of experimental observations by Perrone (1976) and cited by Kerman. He concluded that the two observed regions of ambient noise wind speed dependency indicated the presence of two sound source generation mechanisms or one mechanism that changes sensitivity. Nevertheless, since breaking waves are known to produce bubbles, spray, splash, and turbulence, combinations of these mechanisms may explain the production of sound at frequencies above 500 Hz. Kerman also observed variability in the low-frequency region below 500 Hz. The low-frequency measurements by Whittenborn (1976) with a vertical array of hydrophones above and below the critical depth of the sound channel also showed a dramatic noise level increase associated with breaking waves.

Naval laboratory researchers considered omnidirectional ambient noise a closed subject and the emphasis was on the statistical characterization of noise observed with directional arrays. Ambient noise was simply unwanted random signals to be discriminated. Knowledge of the statistical properties of this noise was required to determine sonar performance, and Dyer (1970) established a fundamental statistical analysis of the shipping component. Following his approach, Wagstaff (1978) conducted experiments with arrays and characterized the beam noise cumulative distribution functions and its persistent directionality. During this period, ambient noise codes were developed that used archival oceanographic and bathymetric data, range-averaged transmission loss, and the uncorrelated plane wave assumption to estimate the omnidirectional noise and directional array response by means of convolution. The response of an array of hydrophones in the noise field was known to be determined by the space-time correlation properties of the field and hydrophone separation. The difficulty in performing large array experiments and the requisite processing analysis necessitated the use of theoretical treatments and simplified analytical models. Cron and Sherman (1962) developed analytical expressions for these correlation functions assuming ergodic random noise sources for volumetric (isotropic) generated noise and surface-generated noise for directional sources. They recognized the noise field was composed of multiple frequency-dependent components such as distant shipping-generated and local wind-generated noise. General agreement with the simplified analytical treatments was found at the higher (above 400 Hz) frequencies. Cox (1973) examined the correlative properties of temporally stationary and spatially homogeneous (ergodic) noise fields. He employed spherical harmonics and their series expansion to describe the cross-spectral density between two sensors and its wavenumber projection. This formulation was found to agree with experimental measurements and to be useful as a basis for optimal array design. These analytical approaches were necessary since acquisition, processing, and beamformer implementation were largely analog. From the mid

1970s to the early 1990s experimental arrays were constructed and experiments conducted to determine the directional response of these systems in the multicomponent noise field. Carey et al. (1997) published results from the early 1980s featuring the response of horizontal arrays in the ambient noise fields of the Mediterranean. The importance of the evolution from analog systems with wide-band tape recorders, to hybrid analog/digital systems with high-density tape recorders, to digital systems and acquisition is central to the development of the understanding of the directional noise field and its correlative properties.

From 1980 to the early 1990s, renewed research focused on the source mechanisms responsible for ambient noise. Between the 1985 Acoustical Society of America meeting in Memphis Farwell 1985) and the 1997 Sea Surface Sound meeting (Leighton 1997), there were six major conferences stressing recent Russian, European, American, and Chinese work (see "Ambient Noise Reviews and Proceedings"). Important new experimental results were reported concerning the sources of noise and the higher-order properties of the noise fields in deep water, shallow water, the Arctic, the Southern Hemisphere, and coastal Asian waters. Breaking waves and the production of splash, microbubbles and clouds were found to have an important role in ambient noise production as well as the scattering of sound from the composite sea surface. The quantification of rain noise was accomplished theoretically and experimentally.

Naval research was directed toward the accurate large-scale noise-field computations to numerically determine the limit ambient noise places on array performance and underwater communication. In addition, a societal interest in the ambient noise background to which marine mammals are exposed developed. This societal interest stresses the importance of accurately summarizing historical as well as current measurements of oceanic ambient noise, especially the contribution of shipping.

The ambient noise schematic produced by Wenz (1962) (Fig. 1.1) provided a qualitative overview of the frequency-dependent ambient noise omnidirectional levels. This conceptual classification by mechanism, frequency, and Beaufort wind force focused much of subsequent ambient noise research along defined frequency ranges and mechanisms.

Even though this paper and schematic have been widely cited and are indeed descriptive of the qualitative ambient noise field, much progress has been made. For example, in the range of frequencies less than 10 Hz, measurements that agree with the theory of the microseismic noise have been made; in the frequency range from 10 Hz to 1 kHz, the non-wind-dependent noise of Wenz and Knudsen has been replaced by the role of shipping and wind-dependent breaking waves; in the frequency region greater than 1 kHz, the roles of bubbles, spray, splash, and rain have been placed on a quantitative basis.

It is important to note that Wenz wisely used the Beaufort wind force as the metric of ambient noise since it not only includes the 10-m wind speed but also includes the appearance of the sea itself. Investigators largely ignored the role of the atmospheric and near-surface boundary layer after his paper, and gradually wind speed alone became the metric. An important contribution would be the quantification of the environmental variables required to be an integral part of future measurements

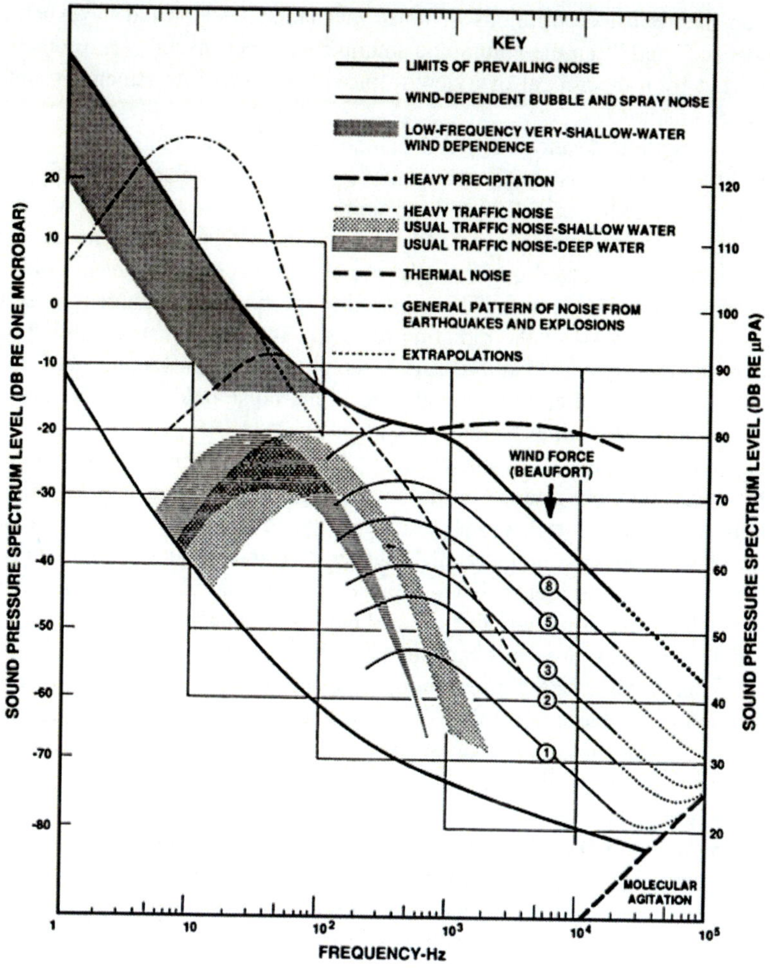

Fig. 1.1 The Wenz curves

necessary to provide quantitative analysis and comparison of experimental results. This is the reason Chapter 2 is on the air–sea interaction zone. The goal was to present a minimal set of parameters to characterize noise measurements. These are the sea state, the Richardson number, and the Reynolds number. It may be possible to use these parameters and satellite observations to provide appropriate environmental knowledge with subsurface noise measurements.

In summary, the literature on this subject is voluminous and beyond the scope of talented scientists to read, understand, and apply. This statement is made in light of the considerable number of and good reviews on this subject of oceanic ambient noise. The questions posed by this state of the art are: "Can a framework be developed for oceanic ambient noise equivalent to the Wenz curve, but based on theoretical descriptions of noise mechanisms supported by experimental evidence?" and "Can the voluminous number of experimental measurements be summarized by a theoretical overview to provide a concise treatment of ambient noise mechanisms?"

This monograph develops a physical understanding of ambient noise mechanisms by coupling analytical treatments of these mechanisms and supporting experimental evidence in a Wenz-like framework of oceanic ambient noise. The air–sea boundary interaction zone and the environmental variables necessary for characterization are treated qualitatively and shown to be important in noise production due to breaking waves. The coherent properties of the ambient noise field are treated on the basis of the radiation from specific mechanisms coupled to the boundary interaction zone. On the basis of the physical realizable mechanisms, the array response, vertical and horizontal, directional noise field characteristics from basin-scale numerical computations presented.

The natural question to be answered is: "What is different in this monograph compared with previous reviews or those works found in conference proceedings?". Certainly an updated Urick (1984) summary would be valuable and necessarily used in developing this monograph. However, this monograph presents not simply a summary of the evidence, but also presents the natural physical-acoustic noise source mechanisms along with the selected but representative experimental results. The ambient noise source mechanisms provide a natural, theoretical framework to discuss and summarize ambient noise measurements. In addition, an example computation of the basin noise field due to surface-generated noise and distributed shipping illustrates the response of an array to the resultant three-dimensional noise field and explains many observed directional noise field characteristics.

The monograph is composed of six chapters followed by ten appendices. The first part of the monograph is written for the acoustical-oceanographer graduate student and researcher. The chapters in this part represent a readable overview of ambient noise measurements and theory focused on natural physical mechanism of noise production. The original monograph was restricted to the archival literature prior to 2000. However, on the recommendation of a reviewer, Chapter 6 was included to discuss contemporary issues in ambient noise research. The importance of bioacoustics and its effects on marine mammals can be found in Richardson and Greene (1995), a detailed treatment of noise impacts, and in Frisk and Bradley (2003), perceived research required to quantify noise impacts

Investigators and students may also be interested in the mathematical basis for many of the phenomena discussed in the first five chapters. These details and derivations are found in the appendices and are included to compensate for the wide-ranging backgrounds of ambient noise researchers. Each appendix contains a separate derivation with appropriate references followed by a summary plate (following the lead of K. Ingard).

For example, the thermal noise limit of the ambient noise simply results from the agitation of the water molecules. The importance of thermal noise to an ambient noise measurement system is first determined by the mean square pressure fluctuation in the water itself and second by the resistive components of the hydrophone amplifier system. The limit of thermal noise is shown in Fig. 1.1. In Appendix F, the thermal noise spectrum is derived by the combination of elemental statistical concepts of the energy density per mode and the density of normal modes in a volume of the ocean. These concepts are related to the literature from statistical physics to room acoustics. This noise places a limit on the detection and measurement of signals in the sea and the minimal detectable plane wave; the result of Mellen (1952) is derived.

Most of the literature on fundamental mechanisms starts with an integral representation of the sources of sound. In Appendix A, these integrals are derived, including the effect of the sea surface (see Appendix G). The application of these source integrals is presented in Chapter 2, where the physical meaning of each is discussed. However, the details are found in the appendices to guide the reader through a general theoretical treatment that encompasses treatments found in the literature.

Appendix B develops from national standards the requisite and the correct quantification of measurements. This appendix was included because there appears to be confusion with the measured noise and its correct reference units. The use of the International System of Units (Système International d'Unités) with logarithmic scales such as the decibel can clarify noise levels in the sea and the societal concern with aquatic life. In short, this appendix answers the question: "What is a deci-bel (decibel)?" Another issue is the correct specification of spectral density and the arcane and inappropriate use of $\sqrt{Hz}$. In preparing this monograph, few changes were made to published results; rather the original figure ordinate and abscissa labels were retained and in several cases clarification is included in the captions.

Sound radiation from splash, that from drops, that from bubbles, and that from bubble clouds are key noise mechanisms at the sea surface. In Appendices C–E and H, suitable expressions are derived to perform calculations and to guide experiments. The result used in the description of these sounds is that they are monopole oscillators below the sea surface, doublets, or point dipoles in the surface. These forms provide a basis for calculations especially when sea surface roughness, sea state, and subsurface microbubble layers are accounted for.

The final two appendices, Appendices I and J, form the basis for computation of the wind-driven noise field and are included to demonstrate the differences in source level determination and representation in computerized codes.

References and Suggested Readings

Arase, E. M. and T. Arase (1965). "Correlation of ambient sea noise." J. Acoust. Soc. Am. 40(1): 205–210.

Carey, W. M., J. W. Reese, et al. (1997). "Mid-frequency measurements of array signal and noise characteristics." IEEE J. Ocean. Eng. 22(3): 548–565.

Carey, W. M. and R. A. Wagstaff (1986). "Low-frequency noise fields." J. Acoust. Soc. Am. 80(5): 1522–1526.

Cox, H. (1973). "Spatial correlation in arbitrary noise fields with application to ambient sea noise." J. Acoust. Soc. Am. 54(5): 1289–1301.

Cron, B. F., B. C. Hassell, et al. (1965). "Comparison of theoretical and experimental values of spatial correlation." J. Acoust. Soc. Am. 37(3): 523–529.
Cron, B. F. and C. H. Sherman (1962). "Spatial-correlation functions for various noise models." J. Acoust. Soc. Am. 34(11): 1732–1736.
Dyer, I. (1970). "Statistics of sound propagation in the ocean." J. Acoust. Soc. Am. 48: 337–345.
Garabed, E. P. and R. A. Finkelman (2005). "Measured vertical noise directionality at five sites in the Western North Atlantic." IEEE J. Ocean. Eng. 30(2) (Special Issue, Archival Papers): 282–285.
Kerman, B. R. (1984). "Underwater sound generation by breaking waves." J. Acoust. Soc. Am. 75(1): 149–165.
Knudsen, V. O., et. al. (1948). "Underwater ambient noise." J. Marine Res. VII(3): 410–429.
Marshall, S. W. (2005). "Depth dependence of ambient noise." "IEEE J. Ocean. Eng. 30(2) (Special Issue, Archival Papers): 275–281.
Mellen, R. H. (1952). "The thermal-noise limit in the detection of underwater signals." J. Acoust. Soc. Am. 24(5): 478–480.
Nichols, R. H. (2005). "Some notable noises: Monsters and machines." IEEE J. Ocean. Eng. 30(2) (Special Issue, Archival Papers): 248–256.
Perrone, A. (1976). *Summary of a One Year Ambient Noise Measurement Program off Bermuda.* Naval Underwater Systems Center, New London Laboratory, NUSC T.R. 4979 (Available DTIC).
Ross, D. (2005). "Ship sources of ambient noise." IEEE J. Ocean. Eng. 30(2) (Special Issue, Archival Papers): 257–261.
Urick, R. J. (1984). Ambient Noise in the Sea, Undersea Warfare Technology Office, NAVSEA, D.O.N. Washington, DC (Available Peninsula Publishing, Los Altos, CA).
Urick, R. J. and A. W. Pryce (1954). *A Summary of Underwater Acoustic Data , Part V, Background Noise.* The Office of Naval Research, Arlington,VA.
Wagstaff, R. A. (1978). "Interactive techniques for ambient noise horizontal directionality." J. Acoust. Soc. Am. 63: 863–869.
Walker, R. A. (1963a). Some widespread high-level underwater noise pulses of apparent biological origin off Cape Cod. In *Marine Bio-Acoustics*, W. N. Tavolga (Ed.), Pergamon Press, New York, NY, 1964, pp. 121–123.
Walker, R. A. (1963b). "Some widespread high-level underwater noise pulses of apparent biological origin off Cape Cod." J. Acoust. Soc. Am. 35: 1816.
Walkinshaw, H. M. (2005). "Measurement of ambient noise spectra in the south Norwegian sea." IEEE J. Ocean. Eng. 30(2) (Special Issue, Archival Papers): 262–266.
Wenz, G. M. (1962). "Acoustic ambient noise in the ocean: spectra and sources." J. Acoust. Soc. Am. 34(12) (Special Issue, Archival Papers): 1936–1956.
Whittenborn, A. F. (1976). Depth dependence of noise resulting from ship traffic and wind, Tracor Corporation, Alexandria, VA, Rpt. T76RV5060 DTIC (AD00692) [Shooter, J. A., T. E. DeMary, and A. F. Whittenborn (1990). IEEE J. Ocean. Eng. 15(4) (Archival Paper): 292–298.

Ambient Noise Reviews and Proceedings

Buckingham, M. J. and J. R. Potter (Eds.) (1995). *Sea Surface Sound '94.* World Scientific Pub. Co. Pte. Ltd., Singapore.
Carey, W. M. (2005). "Special issue on archival papers." IEEE J. Ocean. Eng. 30(2): 245–247.
Carey, W. M. (2010). "Oceanic Noise: Mechanisms, radiation characteristics, and array results. Invited Paper 159th Meeting of the Acoustical Society of America." POMA, J. Acoust. Soc. Am.
Carey, W. M. and E. C. Monahan (1990). "Special issue on sea surface generated noise: 20–2000 Hz." IEEE J. Ocean. Eng. 15(4): 265–266.

Farwell, R. (1985). "Special session air-sea interaction and noise, 110 Meeting of the Acoustical Society of America." J. Acoust. Soc. Am. 78(S1).

Frisk, G., D. Bradley, et al. (2003). *Ocean Noise and Marine Mammals*. Ocean Studies Board, National Research Council, National Academy Press, Washington, DC.

Hersey, J. B. (Ed.) (1974). *International Workshop on Low Frequency Propagation and Noise, Vol. I, Vol. 2*. The Maury Center of Ocean Sciences, Office of Naval Research, Arlington, VA.

Hills, D. (1968). *Sound in the Sea*. Rochester, Capital Records for Marine Resources Inc., Electro Marine Sciences Div., Northridge, CA.

Kerman, B. R. (Ed.) (1988). *Sea Surface Sound: Natural Mechanisms of Surface Generated Noise in the Ocean 15–19 June 1987*. Kluwer Academic Publishers, Dordrecht, The Netherlands.

Kerman, B. R. (Ed.) (1993). *Sea Surface Sound: Natural Mechanisms of Underwater Sound, Sea Surface Sound (2), Cambridge, UK, July 1990*. Kluwer Academic Publishers, Dordrecht, The Netherlands.

Leighton, T. G. (Ed.) (1997). *Natural Physical Processes Associated with Sea Surface Sound*. University of Southampton, Highfield, Southampton.

Medwin, H. (1989). "Special session, sea surface sound, 117 Meeting of the Acoustical Society of America." J. Acoust. Soc. Am. 75(S1).

Richardson, W. J., C. R. Greene, et al. (1995). *Marine Mammals and Noise*. Academic Press, San Diego, CA.

Tavolga, W. N. (Ed.) (1964). *Marine Bio-Acoustics*. Pergamon Press, New York, NY.

Tavolga, W. N. (1965). Review of Marine Bio-Acoustics, State of the Art: 1964, Tech. Rpt.: NAVTRADEVCEN 1212-1. Naval Train Device Center, Port Washington, NY.

Urick, R. J. (1984). Ambient Noise in the Sea, Undersea Warfare Technology Office, NAVSEA, D.O.N., Washington, DC (Also available from Peninsula Publishing, Los Altos, CA, 1986.).

Wenz, G. M. (1972). "Review of underwater acoustics research: Noise." J. Acoust. Soc. Am. 51(2 pt 3): 1010–1024.

Chapter 2
The Air–Sea Boundary Interaction Zone

The research problem addressed by ambient noise investigators since the 1950s is the quantitative determination of the sources of sound in the sea. Investigators quickly realized the importance of intermittent sources of sound, compared with the persistent ambient background. Biological noises and nonbiological (including man-made sounds) sources were considered intermittent and predicated on the areas of operation of naval sonar systems. Certainly one would observe a cacophony of noises, grunts, moans, chirps, etc. in shallow water and arctic areas. The crustaceans (shellfish, shrimp), marine mammals (whales, killer whales, and dolphins) and many fish (croakers) produce loud sounds and these can often dominate the ambient noise background (see the suggested bioacoustic references in Chapter 1). Nonbiological sources such as atmospheric storms, seismic disturbances, and the activities of man such as fishing, offshore oilrigs, and airgun surveys have been studied and are also important contributors to the noise field. However, this monograph focuses on the *natural physical mechanisms* of ambient noise that determine the persistent ambient noise background and the properties of the air–sea interaction zone that determine the characteristics of this sound.

The sea–surface interaction zone (Fig. 2.1) is characterized by the wave spectrum, an atmospheric boundary layer, and a subsurface boundary layer. The atmospheric boundary layer, the marine layer, depends on the roughness of the surface determined by the sea state spectrum, thermal stability, humidity flux, and wind speed. The subsurface layer depends on the thermal stability, suborbital wave motions, and turbulence below the sea surface but also on the presence of microbubble layers and clouds. Indeed, this complex situation is difficult to characterize experimentally because of lack of knowledge of the boundary layer characteristics, which are difficult to characterize theoretically. Nevertheless, a qualitative description of this interaction zone is possible.

As shown in Fig. 2.1, the interaction zone basically is composed of two-phase turbulent layers: spray splash and air above, with bubble clouds, critters, and water below. The general problem for the air–sea layer is the characterization of the state of the sea and the velocity profile above the rough moving sea surface. The problem in the subsurface layer is the characterization of the convection and the presence of microbubbles as a function of sea state and water column stability.

W.M. Carey, R.B. Evans, *Ocean Ambient Noise*, Underwater Acoustics,
DOI 10.1007/978-1-4419-7832-5_2, © William M. Carey and Richard B. Evans 2011

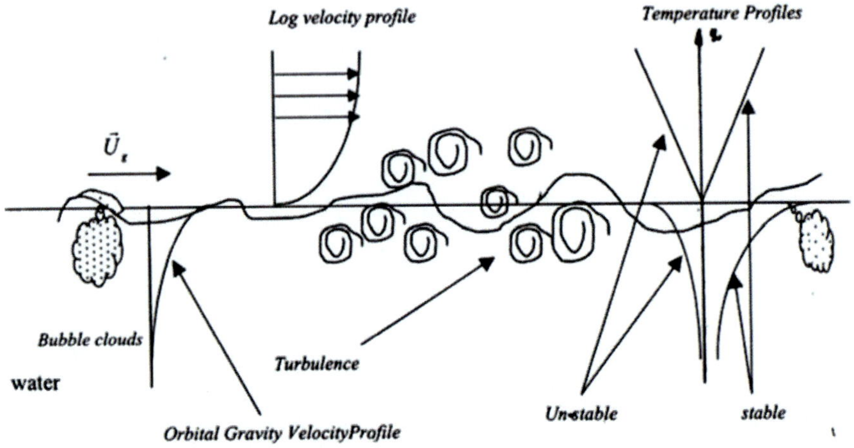

Fig. 2.1 The air–sea interaction zone

The Marine Boundary Layer

The state of the sea has long been a subject of interest for the mariner. Table 2.1 shows the Beaufort scale along with the Hydrographic Office and international scales. Bowditch's *The American Practical Navigator* (Bowditch (1966)) has been widely used in this regard. A key feature of the Beaufort scale is the combination of the visual appearance of the sea surface as well as the wind speed. The visual observations reflect the understanding that the sea state spectrum, fetch, and various near-surface conditions can have a drastic effect on the real state of the sea. Needless to say, the judgment as to the state of the sea can vary from one observer to the next and consequently one would prefer a standard measurement such as wind speed, water temperature, air temperature, or humidity.

As stated previously, the air–sea interaction zone is composed of two turbulent layers, each layer containing multiple unique features. The basic question is what simplification can be made to characterize the complex zone to adequately parameterize the production of sound; can an analytical model with measured parameters describe the state of this zone and parameterize the production of sound? Wenz in his classic paper wisely chose the Beaufort scale (Table 2.1) with its reliance on wind speed, wave height, and appearance of the sea surface, as the parameterization. This choice incorporates the combined effects of mass, momentum, and heat transfer.

One choice is the selection based on empirical evidence of the logarithmic velocity profile of the marine surface layer. Would a measurement of the wind speed at a reference height, 10 m, and a logarithmic velocity profile be an adequate parameterization of this complicated interaction zone? Experience shows to first order that wind speed is a good descriptor and is widely used; but could the exclusive use of the wind speed descriptor also account for much of the noise variability observed?

Table 2.1 The Beaufort scale

Beaufort number	Wind speed: knots	Wind speed: mph	Wind speed: meters per second	Wind speed: km per hour	Seaman's term	World Meteorological Organization (1964)	Estimating wind speed: Effects observed at sea	Estimating wind speed: Effects observed on land
0	under 1	under 1	0.0–0.2	under 1	Calm	Calm	Sea like mirror.	Calm; smoke rises vertically.
1	1–3	1–3	0.3–1.5	1–5	Light air	Light air	Ripples with appearance of scales; no foam crests.	Smoke drift indicates wind direction; vanes do not move.
2	4–6	4–7	1.6–3.3	6–11	Light breeze	Light breeze	Small wavelets; crests of glassy appearance, not breaking.	Wind felt on face; leaves rustle; vanes begin to move.
3	7–10	8–12	3.4–5.4	12–19	Gentle breeze	Gentle breeze	Large wavelets; crests begin to break; scattered whitecaps.	Leaves, small twigs in constant motion; light flags extended.
4	11–16	13–18	5.5–7.9	20–28	Moderate breeze	Moderate breeze	Small waves, becoming longer; numerous whitecaps.	Dust, leaves, and loose paper raised up; small branches move.
5	17–21	19–24	8.0–10.7	29–38	Fresh breeze	Fresh breeze	Moderate waves, taking longer form; many whitecaps; some spray.	Small trees in leaf begin to sway.
6	22–27	25–31	10.8–13.8	39–49	Strong breeze	Strong breeze	Larger waves forming; whitecaps everywhere; more spray.	Larger branches of trees in motion; whistling heard in wires.
7	28–33	32–38	13.9–17.1	50–61	Moderate gale	Near gale	Sea heaps up; white foam from breaking waves begins to be blown in streaks.	Whole trees in motion; resistance felt in walking against wind.
8	34–40	39–46	17.2–20.7	62–74	Fresh gale	Gale	Moderately high waves of greater length; edges of crests begin to break into spindrift; foam is blown in well-marked streaks.	Twigs and small branches broken off trees; progress generally impeded.
9	41–47	47–54	20.8–24.4	75–88	Strong gale	Strong gale	High waves; sea begins to roll; dense streaks of foam; spray may reduce visibility.	Slight structural damage occurs; slate blown from roofs.
10	48–55	55–63	24.5–28.4	89–102	Whole gale	Storm	Very high waves with overhanging crests; sea takes white appearance as foam is blown in very dense streaks; rolling is heavy and visibility reduced.	Seldom experienced on land; trees broken or uprooted; considerable structural damage occurs.
11	56–63	64–72	28.5–32.6	103–117	Storm	Violent storm	Exceptionally high waves; sea covered with white foam patches; visibility still more reduced.	Very rarely experienced on land; usually accompanied by widespread damage.
12	64–71	73–82	32.7–36.9	118–133	Hurricane	Hurricane	Air filled with foam; sea completely white with driving spray; visibility greatly reduced.	
13	72–80	83–92	37.0–41.4	134–149				
14	81–89	93–103	41.5–46.1	150–166				
15	90–99	104–114	46.2–50.9	167–183				
16	100–108	115–125	51.0–56.0	184–201				
17	109–118	126–136	56.1–61.2	202–220				

To examine this question, a discussion of the logarithmic profile and its application to the marine boundary is required.

The Viscous Sublayer

Flow over a smooth plate requires the velocity of the fluid to be zero at the surface of the plate. The change in average fluid velocity, $\bar{u}(z)$, with distance from the plate must be determined by the tangential shear stress. Newton's law of molecular viscosity states

$$\tau = \mu\, d\bar{u}(z)/dz = \rho\,\nu\, d\bar{u}(z)/dz. \qquad (1)$$

At the interface this stress is referred to as the wall shear stress, τ_w, and one can readily see by use of a Taylor series since $\bar{u}(o) = 0$ that

$$\bar{u}(z) = \bar{u}(o) + (\partial\bar{u}/\partial z)_o z + \ldots\ldots \approx (\partial\bar{u}/\partial z)_o z = (\tau_w/\rho)(z/\nu) = u_*^2\,\delta_\nu. \qquad (2)$$

In this expression u_* is the friction velocity at the surface and δ_ν is the thickness of the viscous sublayer. The relative importance of this viscous sublayer to the marine boundary layer can be determined by use of the Reynolds number (R_e), the ratio of the inertial forces (mechanical turbulence in the layer) to the viscous forces. This number can be expressed as $R_e = L_c\bar{u}/\nu$, where L_c is the height of the marine layer (about 10–50 m), $\bar{u}$ is the mean velocity of the air at a distance of (10 m) from the air–sea interface, and ν is the fluid viscosity (about 0.002 m^2/s); the resulting Reynolds number is 5×10^4, indicating turbulent flow. Since the corresponding viscous boundary layer thickness is (10^{-3} m) much less than L_c, simple flat plate theory will by itself not be useful in describing the marine layer. However, the presence of mechanical, buoyancy, heat transport and mass transport, and sea surface motion effects can alter the near-surface profile.

Mechanical Turbulence

One may account for the mechanical turbulence by using a coefficient of eddy viscosity, K_m, and treating the region between a reference distance near the interface, z_o, and the observed height of the turbulent layer.

$$\tau = \rho\,(\nu + K_m)d\bar{u}(z)/dz \cong \rho\, K_m d\bar{u}(z)/dz \qquad z_o < z < L_c \qquad (3)$$

The resulting shear stress at the z_o reference condition becomes $\tau_o = \rho\, K_m(\partial\bar{u}/\partial z)_{z_o}$.

The coefficient of eddy viscosity is known as the "austausch" or exchange coefficient, $A = \rho\, K_m$. Since the goal is to find the velocity profile, observe that $d\bar{u}/dz$ depends on the parameters ν, z, ρ, and τ_o. The π theorem states that with these five dimensional parameters and three fundamental dimensions, two nondimensional ratios can be derived as follows:

$$\begin{aligned} &u_* = \sqrt{(\tau_o/\rho)},\ \textit{the friction velocity and the nondimensional ratios} \\ &\qquad (d\bar{u}/dz)(z/u_*)\ \ \textit{and}\ \ zu_*/\nu. \\ &\textit{Dimensionless analysis yields}\ f_o((d\bar{u}/dz)(z/u_*), zu_*/\nu) = 0 \\ &\qquad \textit{or}\ \ d\bar{u}/dz = (u_*/z)f_1(zu_*/\nu). \end{aligned} \tag{4}$$

The friction velocity, u_*, is necessary to determine the velocity profile. When the distance to the interface is small, such that all roughness elements are less than the reference distance $z_{o\nu}$, (see Fig. 2.2), the viscous effects determine the profile, and R_e is of order 100, then $z_{o\nu}$ is of order 1 mm and can only represent a smooth surface or completely still water. When the surface roughness is much larger than $z_{o\nu}$, z_o is chosen sufficiently large enough to contain the surface roughness, as shown in Fig. 2.2; the larger-scale mechanical eddies dominate and the quantity f_1 needs to be determined. Recognizing the weak dependence of f_1 on zu_*/ν when the reference distance (z_o) is larger than the roughness (h_s) of the interface, one takes the f_1 function to be a constant, $1/\kappa$, where κ is von Kármán's constant. It then follows that

$$\begin{aligned} &d\bar{u}/dz = (u_*/z)(1/\kappa) \\ &\rightarrow\ \ u(z) \cong (u_*/\kappa)ln(z/z_o);\ \ \bar{u}(z_o) = 0\quad h_s < z_o < z < L_c \end{aligned} \tag{5}$$

In Eq. (5), the no-slip condition has been applied at the reference distance, $\bar{u}(z_o) = 0$, with $z_o > h_s$, the roughness distance. With a slight modification of the logarithmic argument, $((z-h_s)/h_s)$, the no-slip condition can be applied at the actual interface, but the distance, z_o, is small compared with the height of the marine layer and this modification has no practical importance.

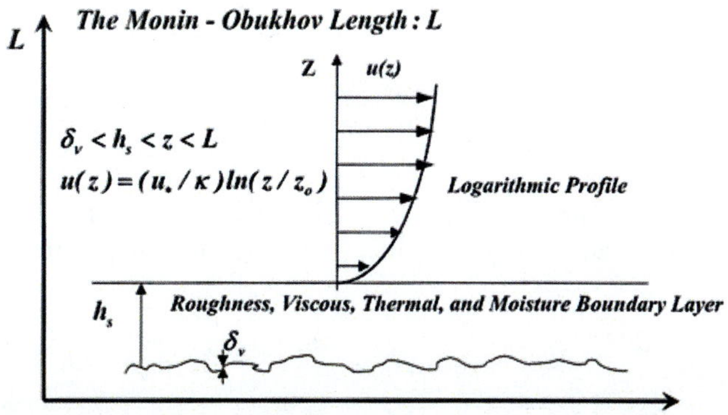

Fig. 2.2 The marine boundary layer for turbulent flow over a rough sea surface

The Effect of Buoyancy

The viscous sublayer and the mechanical turbulence discussed thus far should be augmented by the incorporation of buoyancy effects. Mechanical turbulence is, by its nature, adiabatic, whereas buoyancy, by its nature, is diabatic and dependent on the vertical temperature variation, the lapse rate. When an air parcel is vertically displaced adiabatically, its volume will change in agreement with the ideal gas law, the adiabatic lapse rate. The buoyancy force for such a displacement is the difference between the parcel mass, M_p, and the displaced mass, M_a, times the gravitational acceleration (g):

$$\begin{aligned} &F_b = g(M_a - M_p) = gV(\rho_a - \rho_p) = a_b M_p \\ &\textit{where } \rho_a, \rho_p \text{ are the densities and } p/\rho = RT. \\ &\rightarrow a_b = g(\rho_a - \rho_p)/\rho_p = g(T_p - T_a)/T_a \end{aligned} \tag{6}$$

The buoyancy force yields buoyancy acceleration, a_b, the sign of which determines whether the force is positive, that is, upward, or negative, that is, downward. The adiabatic lapse rate, Γ, is thus the temperature change required by the decrease in pressure to maintain neutral buoyancy. The diabatic lapse rate, γ, is determined by the change of temperature resulting from volume change and heat exchange with the surrounding air. Expanding the temperatures in the above expression in a Taylor series about an equilibrium condition gives

$$a_b = g(\partial T_p/\partial z - \partial T_a/\partial z)\Delta z/T_a = g(\gamma - \Gamma)\Delta z/T_a. \tag{7}$$

This equation shows the importance of the adiabatic lapse rate, Γ; when one has an adiabatic condition $\gamma = \Gamma$, a characteristic of the temperature stratification. When a diabatic lapse rate, γ, exists and normally it does, the buoyancy force can strongly affect the turbulence that occurs in the atmosphere when the production of turbulent energy by the wind stress is just large enough to counter the consumption by the buoyancy force.

This ratio of the consumption of turbulent energy by the buoyancy force to the production of turbulent energy by the wind stress is the Richardson number, R_i. If $\Delta z = 1$, then

$$a_{bd} = g(\gamma - \Gamma)/T_a = (g/T_a)d\Theta/dz \tag{8}$$

and one has four fundamental quantities – $d\bar{u}/dz$, $d\Theta/dz$, g, and T_a – with three fundamental dimensions, so one nondimensional variable can be formed:

$$R_i = (g/T_a) \cdot \frac{d\Theta/dz}{(d\bar{u}/dz)^2} = (g/T_a) \cdot \frac{(\gamma - \Gamma)}{(d\bar{u}/dz)^2} \quad . \tag{9}$$

The importance of the adiabatic lapse rate and the relative importance of the diabatic lapse rate are observed. The quantitative values of the lapse rates are not as important

Table 2.2 Thermal stability

1	$\gamma = -\,dTdz = \Gamma$	$T_p = T_a$	Neutral	The displaced air parcel neither rises nor falls
2	$\gamma > \Gamma$	$T_p > T_a$	Unstable	Vertical displacement is amplified by buoyancy
3	$\gamma \Gamma$	$T_p < T_a$	Stable	Vertical displacement is dampened

as the relative differences. (The dry adiabatic lapse rate in the atmosphere is of order 1°C/100 m.). Three cases of stability can be determined as neutral, unstable, and stable as shown in Table 2.2.

The Influence of Heat Transport

The significant factor for heat transport is the departure of the eddy temperature from the temperature of the surrounding air. Because of this temperature difference, heat transfer can occur and thus eddies can transport heat across the flow. Just as the mechanical turbulence was described by a coefficient of eddy viscosity, K_m, one may also describe the heat transport, q, or flux by a coefficient of eddy heat conduction, K_q, with $q = \rho K_q C_p(\gamma - \Gamma)$, where C_p is the specific heat. The result is a modification of the Richardson number to a flux form:

$$R_f = R_i(K_q/K_m). \tag{10}$$

The number of fundamental quantities above has increased from four (z, z_0, u_*, and κ) with the addition of g, C_p, ρ, q, and T to nine. Dimensionless analysis will produce multiple nondimensional numbers and would be unnecessarily complex for the purpose of this treatment. A further simplification is thus required, such as that proposed by Lettau (1949) and latter by Monin and Obukhov (1953) [references from Slade (1968), Kitaigorodskii (1972), and Plate (1971)].

The Monin–Obukhov Length

As shown in Fig. 2.2, the region of interest to our problem is that between some unspecified distance that encompasses the roughness of the surface, $z_o \geq h_s$, and heights specified by some empirical but characteristic length, L_c. In this region the flow can be considered stratified-parallel-turbulent flow from near the roughness distance to the specified characteristic height with constant stress and heat flux or a region with constant momentum and heat flow. With this assumption, the quantities of interest are g/T_a, u_*, and $K_q/C_p\,\rho$ and the Monin–Obukhov-derived unique length scale is

$$L = [u_*^3/\kappa(g/T_a)][-q_\Theta/C_p\rho]^{-1} = -u_*^3 C_p \rho T_a/\kappa g q_\Theta. \tag{11}$$

The Monin–Obukhov length, L, can be negative for unstable conditions, positive for stable conditions, and become infinite as $\gamma \to \Gamma$. The near-wall region can first be governed by viscous properties and then by additional turbulence by the interaction of the flow and the rough boundary. However, the dynamic layer between this roughness parameter and the characteristic length $L_c = |L|$ because of the constant momentum, heat, and mass flux is a self-similar region with logarithmic profiles. The Richardson number now can be written as

$$\begin{aligned} &R_{if} = R_i K_q / K_m = (u_* Ta/\kappa\, g)(1/Lc(d\bar{u}/dz)) \\ &and \\ &R_i = [(g/\Theta)d\Theta/dz]/[d\bar{u}/dz]^2 = z/L_c \end{aligned} \qquad (12)$$

where $R_i = z/L_c$ is the local Richardson number. Thus, this complex problem (Fig. 2.2) of the marine boundary layer over a rough surface has been treated with a viscous sublayer, δ_ν, less than a roughness layer that contains the influence of viscous, thermal, and moisture sublayers, h_s, and a dynamic layer between this roughness layer and the Monin–Obukhov height. The velocity profile in this dynamic region with constant heat and momentum fluxes was shown to be

$$\begin{aligned} &d\bar{u}(z)/dz = u_*/\kappa z \\ &and \\ &\bar{u}(z) = (u_*/\kappa)\, ln(z/z_o) \quad with \quad \delta_\nu < h_s \le z_o \le z \le L_c\,. \end{aligned} \qquad (13)$$

This development, although not particularly quantitative, should provide a qualitative understanding of effect of the complex interaction of roughness, momentum, and heat transport on the development of the logarithmic profile.

The Combined Influence of Mass, Momentum, and Heat Transport

Although the mass transfer due to moisture has not been included, it is sufficient to state that the laws of viscosity, heat conduction, and diffusion are all similar and for each phenomenon an eddy coefficient can be approximated and treated as has been done thus far. When all three effects, momentum, heat, and mass transfers, are considered, it is convenient to use relative quantities with the change in velocity between a reference height and the roughness height, $\delta\bar{u} = \bar{u}(z_r) - u(z_o)$, the relative temperature difference, $\delta\bar{\Theta} = \bar{\Theta}(z_o) - \bar{\Theta}(z_r)$, and the specific moisture difference, $\delta\bar{m}_e = \bar{m}_e(z_o) - \bar{m}_e(z_r)$. The distance z_r is a reference distance and is usually taken to be a height of 10 m from the sea surface. The quantities of interest at this reference height are the momentum flux, $\rho\bar{u}_r^2$, the heat flux, $\rho C_p \delta\bar{\Theta}\bar{u}_r$, and the moisture-humidity flux, $\rho\delta\bar{m}_e\bar{u}_r$. Their description is based on the coefficients of drag, C_u, heat exchange, C_Θ, and moisture exchange, C_{me}:

$$
\begin{aligned}
&C_u(z_r/h_s,\ z_r/L, h_s/\delta_\nu) = \tau/\rho\bar{u}_r^2 \\
&C_\Theta(z_r/h_s,\ z_r/L, h_s/\delta_\Theta,\ P_{R\Theta}) = q_\Theta/\rho C_p\delta\bar{\Theta}\bar{u}_r \\
&C_{me}(z_r/h_s,\ z_r/L, h_s/\delta_{me},\ P_{Rme}) = w_{me}/\rho\delta\bar{m}_e\bar{u}_r
\end{aligned}
\tag{14}
$$

The three sublayers for the conditions found near the sea surface result in Prandtl numbers ($P_{R\Theta}$ *and* P_{Rme}) of unity, and consequently

$$
P_{R\Theta} = \delta_\nu/\delta_\Theta = \nu/\nu_\Theta = 1, \quad P_{Rme} = \delta_\nu/\delta_{me} = \nu/\nu_{me} = 1. \tag{15}
$$

The result is the viscous, thermal, and moisture sublayers are approximately of equal thickness. These layers and the resulting roughness layer characterize the near-surface vertical transfer of momentum, heat, and moisture. Consequently, the coefficients only depend on three nondimensional ratios of z/h_s, R_i, and R_e. The key result of this analysis can be summarized by use of nondimensionalized velocity, temperature, and moisture:

$$
\tilde{u} = \bar{u}(z)/u_* \,;\ ,\tilde{\Theta} = \bar{\Theta}(z)u_*/(-q_\Theta/\rho C_p),\ \tilde{m}_e = \bar{m}_e\rho u_*/(-w_{me}). \tag{16}
$$

The governing equation is

$$
d\tilde{u}/dz = d\tilde{\Theta}/dz = d\tilde{m}_e/dz = 1/\kappa z \ ; \ \delta_\nu < h_s \cong z_o \le z \le L_c. \tag{17}
$$

The coefficients are

$$
\begin{aligned}
&C_u(z/h_s, R_i, R_{es}) = (ln(z/z_o)/\kappa)^{-2} \\
&C_\Theta = C_u(z/h_s, R_i, R_{es})/(1 + C_u^{1/2}\delta\tilde{\Theta}) \\
&C_{me} = C_u(z/h_s, R_i, R_{es})/(1 + C_u^{1/2}\delta\tilde{m}_e) \\
&C_u \cong C_\Theta \cong C_{me} \qquad z_o \le z \le L_c\ .
\end{aligned}
\tag{18}
$$

The final result is that in the dynamic marine layer these coefficients are approximately equal and are usually referenced to a height of 10 m.

The single factor of importance in the coefficients above besides the friction velocity is the roughness parameter, h_s. When the Reynolds number, $R_e = h_s u_*/\nu = h_s/\delta_\nu$, is large, the roughness parameter, z_o, is proportional to h_s, and the near-wall condition is roughness-controlled. When the Reynolds number is small, then the parameter z_o is comparable to δ_ν and the sea surface is calm and viscous forces dominate. Thus, the sea surface can be described as smooth, incompletely rough, and completely rough. In the completely rough case $z_o = A_s h_s$, where A_s is a constant factor depending on the steepness of the roughness.

The Roughness Scale and Motion of the Surface

This treatment of the marine boundary layer has assumed that the surface is immobile and rigid. The rigid boundary condition may well be justified in light of the large density difference between air and water; however, the sea surface is definitely in motion. The supposition is that a quantitative treatment of the boundary layer is beyond the scope of this monograph, whereas a qualitative description that is useful in the characterization of oceanic ambient noise would be valuable. In this regard two factors are still to be discussed, the roughness and sea surface motion. Here the argument of Kitaigorodskii (1972) is useful. Although his treatment may not lend itself to a quantitative description of the boundary layer, the motion of the sea surface, and its roughness, it provides insight as to the role these various factors have in describing the characteristics of oceanic ambient noise.

The surface can be considered a rough surface with the roughness contained in a layer, the z_o roughness parameter, which is greater than the effective roughness and the viscous boundary layer as shown in Fig. 2.2. If a self-similar turbulence region is maintained up to this distance, the logarithmic velocity profile applies:

$$\bar{u}(z) = (u_*/\kappa)\ln(z/z_o) \quad \text{with} \quad \delta_\nu < h_s \le z_o \le z \le L_c. \tag{19}$$

If the roughness of the surface waves moves downstream with a phase velocity of u_{wc} with an amplitude comparable to h_s, then in a Cartesian coordinate frame moving with the waves at velocity u_{wc}, the turbulent flow will be proportional to the relative motion:

$$\begin{aligned} \bar{u}(z) - u_{cw} &= (u_*/\kappa)\, ln(z/z_o), \quad \delta_\nu < h_s \le z_o \le z \le L_c \\ &\rightarrow u(z)/u_* = (1/\kappa) ln(z/z_o) + u_{cw}/u_* \\ &= (1/\kappa) ln(z/h_s\, exp(-\kappa u_{cw}/u_*)). \end{aligned} \tag{20}$$

Thus, the motion of the boundary in this simple example can be incorporated in the logarithmic velocity profile by the inclusion of a factor $h_s \cdot exp(-\kappa u_{wc}/u_*) >> \delta_\nu$, a modification to the roughness parameter, an effective roughness parameter. This is not an exact solution to the boundary layer of a moving boundary, but states that the moving boundary presents a different near-surface condition because of its motion. If the amplitude h_s is the amplitude of the portion of the sea state spectrum $S(\omega)$ in some interval $d\omega$, then the mean square roughness is

$$\begin{aligned} &h_s^2(k) \cdot exp(-2\kappa u_{cw}/u_*)/2 = S(k)dk\,, \quad u_{cw} = g/\omega \\ &\rightarrow <h_s^2> = 2\int_o^\infty S(k) exp(-2\kappa u_{cw}(k)/u_*)dk \cong <z_o^2> \end{aligned} \tag{21}$$

The result is the simplest form of marine boundary layer including a rough moving surface. Undoubtedly, this equation may not, in general, describe the complicated near-surface condition, but it highlights the parameters that need be considered, such as the sea surface spectrum.

In summary, the logarithmic velocity profile has been applied to this marine layer problem following the approach of Kitaigorodskii (1972) that shows the wind stress coefficient is fundamentally dependent on three quantities:

$$C(z, u) = C(z/h_s, R_i, R_{es}), \tag{22}$$

where z is the observation point, h_s is the characteristic scale of the roughness, R_i is the Richardson number, and R_{es} is the Reynolds number for roughness. Kitaigorodskii also showed, for a moving boundary under neutral stability conditions, that

$$u(z)/u_* \simeq 1/\kappa\, ln(z/h_s) = C(z, u(z))^{-1/2}, \tag{23}$$

where $u_* = \sqrt{(\tau/\rho)}$ is the friction velocity, κ is von Kármán's universal constant, and h_s is given by

$$h_s \simeq [2 \int_0^\infty S(k) exp(-2\kappa u_{cw}(k)/u_*) d\omega]^{1/2} \tag{24}$$

with $S(\omega)$ being the frequency spectrum of the wave field.

In general, the wind stress coefficient is dependent on the stability of the boundary layer (mechanical turbulence, heat and mass transport) as well as the roughness scale modified by the motion. This roughness scale depends on the sea state spectrum, which in turn is a function of time, the wind at a given speed, and the fetch. These fundamental considerations clearly indicate that wind speed alone may not simply be the best indicator of the state of the sea. As will be observed in later sections, satellite and large-scale meteorological observations can be used to correctly parameterize the state of the sea.

A Summary of the Wind Stress Coefficient and Critical Friction Velocity

Boundary layer investigations employ a standard measurement height of 10 m. The wind stress coefficient at this reference point is $C_{10}(U_{10})$ and is related to u_* with use of the previously discussed logarithmic velocity profile. The variables such as wind speed, U_{10}, and Richardson number are also referenced to this height. Amorocho and DeVries (1980) examined many boundary measurements by plotting C_{10} and u_* versus U_{10} for a wide range of wind speed conditions. The results are shown in Fig. 2.3. In general, they found three distinct wind speed regions. The first region is found prior to the onset of wave breaking with C_{10} constant. The second region, labeled as a transition region, is for wind speeds between 7 and 20 m/s. In this region, both u_* and C_{10} have nonlinear dependencies on wind speed. Finally, for wind speeds greater than 20 m/s, there is a saturation region with C_{10} becoming relatively constant.

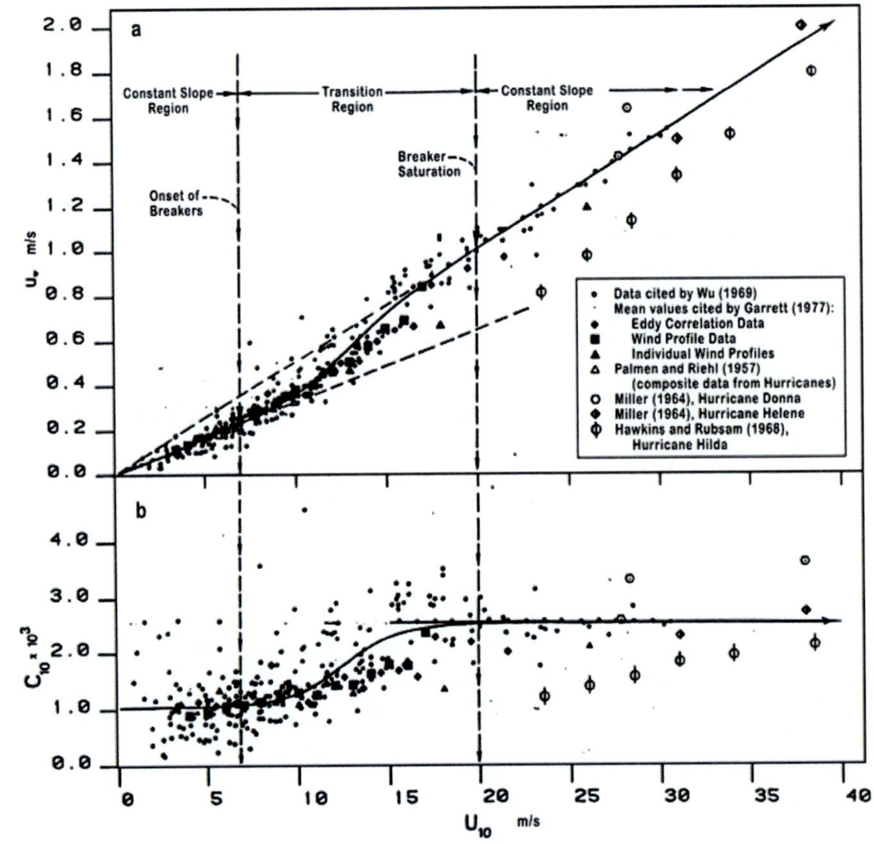

Fig. 2.3 Estimates of the friction velocity and wind stress coefficient from measurements over water surfaces

Friction Velocity and Breaking Waves

An important feature of this summary is the importance of wave breaking. Kerman states that this wave breaking occurs when the kinetic energy of an eddy near the surface, $\rho l_e^3 u_*^2$, can overcome the surface tension of the water, $l_e^2 T_{sur}$. If the limiting acceleration for convective overturning of the crest is taken as $l \approx u_*^2/g$, then a critical friction velocity $u_{c*}^4 = 4\,gT_{sur}/\rho$ can be defined such that when the friction velocity exceeds this value waves begin to break. This expression for the critical friction velocity is also the minimum phase velocity of the gravity–capillary waves. The relationships discussed thus far are qualitative and need not be quantitative for the description of environmental measurements necessary for oceanic ambient noise studies. However, it should be clear at this point that wind speed alone will not suffice. To stress the importance of this wave-breaking phenomenon, Fig. 2.4 shows a breaking wave observed by Su (1984).

Brackish water measurements, such as the experiments by Su (1984) with three-dimensional waves (Fig. 2.4), show a typical sequence of events: from the initial wave breaking; to the formulation of a columnar bubble plume; to a more diffuse cloud swept by the subsurface orbital motion; and finally to dispersion of near-surface bubbly features. Although freshwater experiments may be useful in visualizing the sequence, the bubbly mixture in these types of experiments appears to be composed of larger bubbles, faster rise times, and consequently different acoustic characteristics. The difference in the bubble sizes and rise times also may result in different sound production mechanisms.

Thorpe (1982, 1983, 1986) and his colleagues performed a series of experiments with upward-looking sonar in lakes and in the deep ocean. He observed that bubble clouds were convected to meter depths by wave-breaking turbulence and vorticity with subsequent ordered patterns consistent with Langmuir circulation. Thorpe's results showed an exponential decrease in volume scatter strength with depth, a mean depth of penetration that was proportional to U_{10} and the air–water temperature difference ($\Delta\theta_{AW}$). Different cloud characteristics for stable ($\Delta\theta_{AW} > 0$) and unstable ($\Delta\theta_{AW} < 0$) conditions were also observed. For example, the stable condition was found to result in a "billowy" cloud structure, whereas the unstable condition resulted in a columnar characteristic. This observation is reinforced by the parametric dependence of the logarithmic profile on R_i previously discussed. Observations by Crawford and Farmer (1987) confirmed these breaking-wave effects: the exponential distribution of bubble density with an "e folding" depth between 0.7 and 1.5 m, a bubble density variation with $u_{10}^{3\pm0.3}$, a weak dependence on the Langmuir circulation, and "v"-shaped columnar clouds for ($\Delta\theta_{AW} < 0$).

Thorpe also observed pronounced differences in clouds produced by wave breaking in freshwater and salt water. He attributed these differences to chemical effects discussed by Scott (1975) to explain Monahan's (1971) observation concerning freshwater whitecaps. That is, under nearly identical physical conditions, bubble distributions produced in salt water have a smaller mean radius and a larger number of bubbles. According to Scott (1986), surface chemical effects can be an important factor preventing coalescence, and "significant differences observed in the duration of freshwater and salt water whitecaps may be ascribed to these effects." Bubbles in salt water were greater in number, smaller, more densely packed, carried deeper, and slower to rise to the surface than those formed in freshwater by a similar wave-breaking event.

Pounder (1986) showed a distinct temperature-dependent difference between distilled water (coalescence occurs) and salt water (coalescence does not occur). He attributed this difference to an ionic effect. Pounder's laboratory observations support the conclusion drawn by Scott and Thorpe that microbubble distributions result from salt water wave breaking.

Bubble size measurements [made by Medwin (1977), Kolovayev (1976), Johnson and Cooke (1979), Bouguel (1985), and Crawford and Farmer (1987) and then reviewed by McDaniel (1987) and Carey and Fitzgerald (1990)] had bubble radius distributions between 50 and 70 μm and an exponential numerical decrease

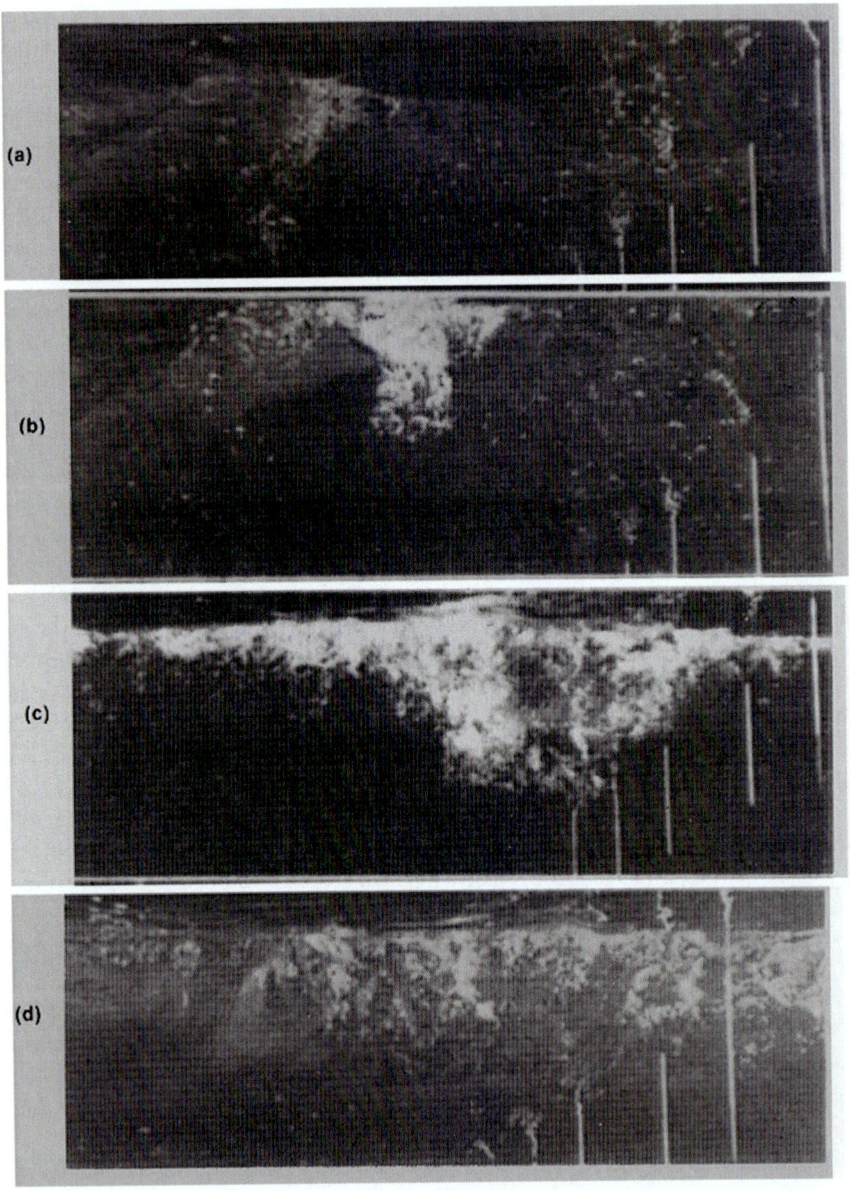

Fig. 2.4 Three-dimensional crescent-shaped breakers that resemble deepwater oceanic breakers are viewed in 2-s intervals from under the breaking wave. The vertical strings are spaced at 30.5-cm intervals and have 15.3 white and black sections. The wave steepness, the initial amplitude a_o times the initial wavenumber k_o, is 0.33. Su (1984)

with an increase in radius. Kolovayev observed that at wind speeds of 13 m/s, all bubbles were less than 350 μm. The distribution maximum was found to shift to larger radii with increasing wind speed and depth.

Monahan (1990) developed a description of the evolution of a bubble plume and cloud from a breaking wave consistent with these results. Shown in Fig. 2.5 are the results of his analysis for a wind speed of 13 m/s. The α plume, first panel, occurs within 1–2 s of the breaking event and its characteristic depth is 0.5 m with a volume fraction of $4 \times 10^{-2} - 8 \times 10^{-2}$ on the basis of the extrapolated bubble size distribution for the β plume. The second panel shows his "β" plume, a bubble size distribution derived from his aerosol generation model, and a bubble size distribution based on the measurements of Johnson and Cooke (1979). The β plume is estimated to have duration between 1 and 10 s and a volume fraction based on an integrated size distribution of $1 \times 10^{-4} - 2 \times 10^{-4}$. In the third panel, the characteristic of the cloud has a bubble size distribution consistent with the measurements of Johnson and Cooke and a scaling based on cloud sea surface ocean to whitecap area of 25:1. The duration of this feature is estimated to be on the order of 100 s and the average volume fraction is between 10^{-6} and 10^{-7}.

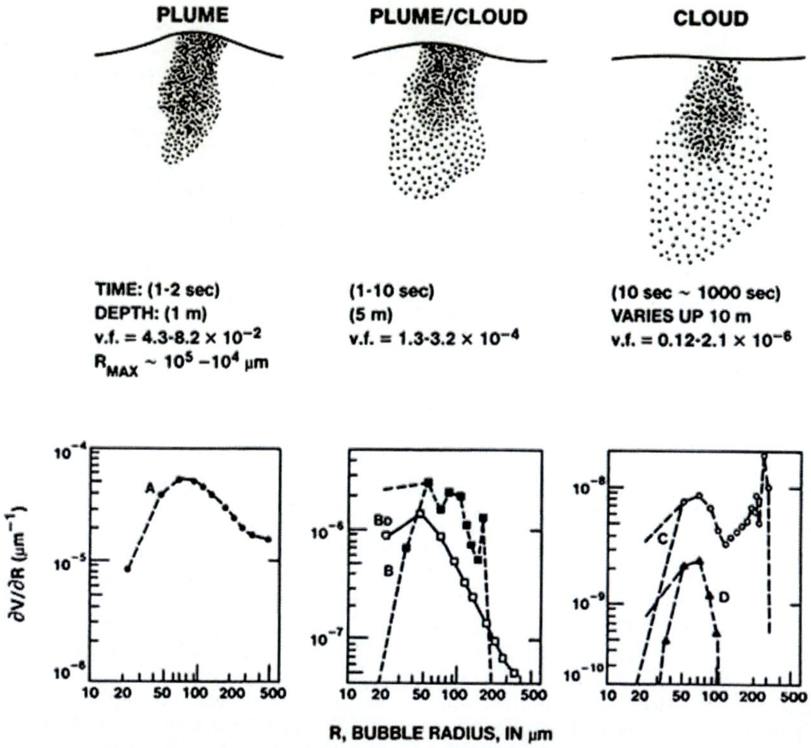

Fig. 2.5 Evolutionary model of a wave-breaking bubble plume and cloud as a sequence of temporal samples covering time periods of 1, 10, and 100 s Monahan (1989)

The measurement of the bubble size distribution, void fraction, and spatial characteristics of bubble clouds and plumes has been shown to be difficult. Nevertheless, general characteristics have emerged concerning breaking waves, the determination of the subsurface characteristics, and the temporal evolution of these features.

Monahan used exponential variation in depth, cross-sectional area, and time with measured "e folding" characteristics to scale the results for the α, β, and γ plumes. These features are shown schematically in Fig. 2.6 as a qualitative picture of the near-surface boundary condition to guide sound scattering and noise measurements.

The surface manifestation of the active-stage class A whitecap (Fig. 2.6) lends itself to observation by shipboard and airborne observation as well as satellite measurement. This whitecap appearance was the basis for the previously discussed

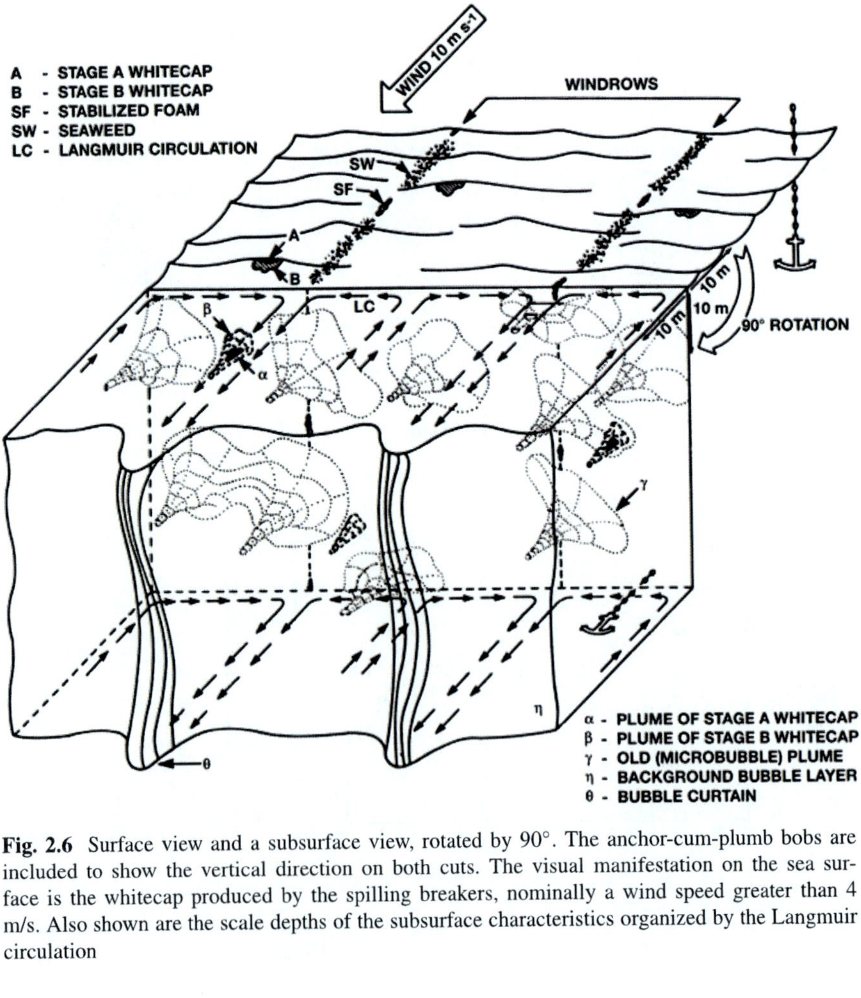

Fig. 2.6 Surface view and a subsurface view, rotated by 90°. The anchor-cum-plumb bobs are included to show the vertical direction on both cuts. The visual manifestation on the sea surface is the whitecap produced by the spilling breakers, nominally a wind speed greater than 4 m/s. Also shown are the scale depths of the subsurface characteristics organized by the Langmuir circulation

Beaufort scale used by Wenz. Satellite-sensed microwave backscatter may well be a viable means of automatically sensing the sea state by the scattering from the active whitecaps and has been proposed by both Monahan and Kerman. The class A active whitecaps and concentrated α bubble plumes are of short duration and acoustically significant. The second feature shown in this figure is the older class B whitecap, the hazy foam patch that remains after the wave has passed. These features cover a wider area and have a microwave emissivity that can be observed with multichannel microwave radiometers. Monahan (1990) discussed the characteristics in detail. The whitecap coverage and Beaufort scale are related, and combine the heat mass and transport effects into an observable parameter.

The Whitecap Index

Wilson (1983) found that ambient noise levels varied in proportion to the "whitecap" index, W(u), of Ross and Cardone (1974) and proposed three regions of wind speed dependency as shown in Table 2.3.

Table 2.3 The Whitecap index

I	$W(u) = 0$	$u < 4.5$ m/s
II	$W(u) = (4.6 \times 10^{-3})U^3 - (4.9 \times 10^{-2})U^2 + (4.63 \times 10^{-1})U - 1.5$	$4.5 < U < 15$ m/s
III	$W(u) = (20.97)(U/15)1.5$	15 m/s $< U$

Wu (1980) stated that the whitecap index W(u) should be related to the energy flux of the wind under equilibrium conditions. The energy flux ($\dot{E}$), or the rate of doing work, is related to the wind stress (τ) and a surface drift current (V):

$$W(U_{10}) \propto \dot{E} = \tau V \propto \tau U_* \propto C_{10}^{3/2} U_{10}^3, \tag{25}$$

where $W(U_{10})$ is the percentage of the sea surface covered by whitecaps, τ is the shear stress at the surface, U_* is the friction velocity, and U_{10} is the 10-m-elevation wind speed. This relationship between the whitecap index and the wind stress coefficient C_{10} ties the whitecap index to fundamental parameters governing the exchange of momentum, mass, and energy in the sea surface interaction zone.

The whitecap index can be obtained either by measuring the index directly or by estimating the wind stress coefficient from boundary layer measurements. Monahan (1990) [see also O'Murcheartaigh and Monahan (1986), Monahan and O'Murcheartaigh (1981)] determined the wind speed variation of the whitecap index by fitting

$$W(U_{10}) = \alpha U_{10}^{\lambda} \tag{26}$$

to measured fractional whitecap coverage. Monahan found that $\lambda = 3.41$ provided the best fit to all data, with individual sets of data yielding values of λ between 2.55

and 3.75. Monahan also proposed classifying whitecaps as class A, young, and class B, old.

Wu (1981, 1986) contended that $C_{10}(U_{10}) \propto U_{10}^{1/2}$, and consequently a variation of W with $U^{3.75}$. However, Wu (1980) earlier recommended a linear dependence of $C_{10}(U_{10})$ on wind speed. His review of measured wind stress coefficients as a function of U_{10}^{3m} showed most values of m between 1 and 1.3 [also, see Large (1981), Donelan (1982), and Smith (1980)]. The reader is cautioned concerning the use of relationships between the whitecap index and the wind stress coefficient $C_{10}(U_{10})$ such as the qualitative relationships shown in Fig. 2.7 as investigations are continuing and a quantitative relationship has not been developed.

The results shown in Fig. 2.7 are remarkably similar to the acoustic noise level characteristics. Most observations of noise are in the 7–20-m/s wind range, thus a wind speed dependency similar to the variation in wind stress coefficient can be expected. Since the wind stress coefficient depends on the logarithmic profile and, in turn, on the Reynolds and Richardson numbers, the sea state spectrum, and the friction velocity, ambient noise levels may also have these characteristics. Thus,

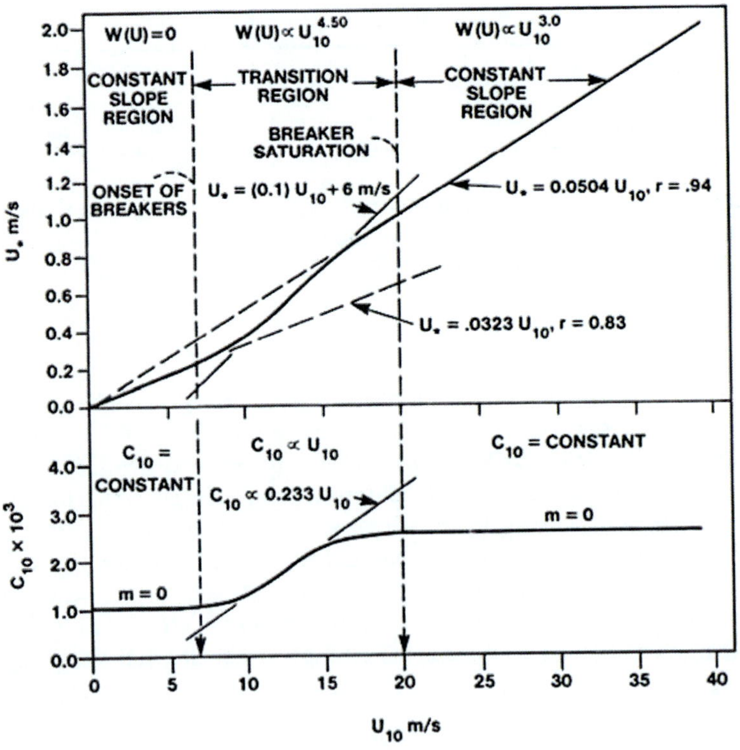

Fig. 2.7 Summary of the relationships for the wind stress coefficient at 10 m, C_{10}, and the estimated friction velocity, u_*, versus the 10-m wind speed, u_{10}, with whitecap indexes, $W(u^m_{10})$, delineated

one may conclude that ambient noise measurements should be performed that concurrently measure the wind speed (U_{10}), the air–water temperature difference as a simplified R_i measure, the sea state spectrum as an indicator of roughness, and the moisture content necessary for correct estimation of R_e. The measurement of whitecaps by the use of satellite observations, satellite estimates of the 10-m wind speed and large-scale meteorological computational tools may be useful in the calculation of wind-induced ambient noise.

References and Suggested Readings

Amorocho, J. and J. J. DeVries (1980). "A new evaluation of the wind stress over water surfaces" J. Geophys. Res. 85(C1): 433–442.

Bouguel, M. and S. Baldy (1985). "Measurement of bubbles in a stationary field of breaking waves by a laser-based single-particle scattering technique." J. Geophys. Res. 90(C1): 1037–1047.

Bowditch, N. (1966). *American Practical Navigator*. U.S. Navy Hydrographic Office, Government Printing Office (H.O. Pub. No. 9), Washington, DC.

Carey, W. M. and J. W. Fitzgerald (1990). Low frequency noise from breaking waves. In *Natural Physical Sources of Underwater Sound*. Cambridge, UK. B. R. Kerman (Ed.), Kluwer Academic Publishers, Dordrecht (1993), pp. 277–304.

Crawford, G. B. and D. M. Farmer (1987). "On the spatial distribution of ocean bubbles." J. Geophys. Res. 92(8): 8231–8243.

Donelan, M. A. (1982). The dependence of Aerodynamic Drag Coefficient on Wave Parameters. Proceedings of the 1st Conference of Meteorology and Air Sea Interaction of the Coastal Zone, American Meteorological Society, pp. 381–387.

Hinze, J. O. (1959). *Turbulence*. McGraw-Hill Book Co., NY, pp. 13–25.

Johnson, B. D. and R. C. Cooke (1979). "Bubble populations and spectra in coastal waters: A photographic approach" J. Geophys. Res. 84(C7): 1038–1042.

Kerman, B. R. (1984). "Underwater sound generation from breaking waves." J. Acoust. Soc. Am. 75(1): 149–164.

Kitaigorodskii, S. A. (1972). *The Physics of Air-Sea Interaction*. U.S. Department of Commerce, NTIS, Springfield, VA.

Kolovayev, P. D. (1976). "Investigation of the concentration and statistical size distribution of wind produced bubbles in the near surface ocean layer." Oceanology 15: 659–661.

Large, W. G. and S. Pond (1981). "Open ocean momentum flux measurement in moderate to strong winds." J. Phys. Oceano. 11: 324–336.

McDaniel, S. T. (1987). Subsurface Bubble Densities from Acoustic Backscatter Data, ARL/TM-87-57, Applied Research Lab, Pennsylvania State University, State College, PA.

Medwin, H. (1977). "In situ acoustic measurements of microbubbles at sea." J. Geophys. Res. 82(6): 971–976.

Monahan, E. C. (1971). "Oceanic whitecaps." J. Phys. Oceanogr. 1: 138–144.

Monahan, E. C. (1988). Whitecap coverage as a remotely monitorable indication of the rate of bubble injection into the oceanic mixed layer. In *Sea Surface Sound*. B. R. Kerman (Ed.), Kluwer Academic Publishers, Boston, pp. 85–86.

Monahan, E. C. and M. Lu (1990). "Acoustically relevant bubble assemblages and their dependence on meteorological parameters." IEEE J. Ocean. Eng. 15(4): 340–349.

Monahan, E. C. and I. O'Murcheartaigh (1981). "Optimal power-law description of oceanic whitecap coverage dependence on wind speed." J. Phys. Ocean. 10: 2094–2099.

Monahan, E. C. and T. Torgersen (1991). The enhancement of air sea gas exchange by oceanic whitecapping. In *Air-Water Mass Transfer*. S. C. Wilhelms and J. S. Gulliver (Eds.), American Society of Civil Engineers, New York, NY, pp. 608–617.

Monahan, E. C. and M. A. Van Patten (Eds.) (1989). *The Climate and Health Implications of Bubble-Mediated Sea-Air Exchange*. Connecticut Sea Grant College Program, University of Connecticut at Avery Point, Groton, CT.

Monahan, E. C. and C. R. Zietlow (1969). "Laboratory comparisons of fresh-water and salt-water whitecaps." J. Geophys. Res. 74: 6961–6966.

O'Murcheartaigh, I. G. and E. C. Monahan (Eds.) (1986). *Oceanic Whitecaps*. D. Reidel Publishing Co., Boston, MA.

Plate, E. J. (1971). *Aerodynamic Characteristics of the Atmospheric Boundary Layers*. U.S. Atomic energy Commission, Available NTIS-TID-25465, National Technical Information Service, U.S. Dept. of Commerce, Springfield, VA.

Pounder, C. (1986). Sodium chloride and water temperature effects on bubbles. In *Oceanic Whitecaps*. E. C. Monahan and G. MacNiocaill (Eds.), Reidel, Dordrecht, Holland, p. 278.

Ross, D. B. and B. Cardone (1974). "Observations of oceanic whitecaps and their relation to remote measurements of surface wind speed." J. Geophys. Res. 79: 444–452.

Scott, J. C. (1975). "The of salt in whitecap persistence." Deep Sea Res. Oceanogr. Abstr. 22(10): 653–654.

Scott, J. C. (1986). The effects of organic films on water surface motions. In *Oceanic Whitecaps*. E. C. Monahan and G. MacNiocaill (Eds.), Reidel, Boston, pp. 159–165.

Slade, D. H. (1968). *Meteorology and Atomic Energy*. U.S. Atomic energy Commission, Available NTIS-TID-24190, National Technical Information Service, U.S. Department of Commerce, Springfield, VA.

Smith, S. D. (1980). "Wind stress and heat flux over the ocean in gale force winds." J. Phys. Oceano. 10: 709–726.

Smith , S. D. (1981). "Comment on "A new evaluation of the wind stress coefficient over water surfaces"." J. Geophys. Res. 86(C5): 4307.

Su, M. Y. (1984). Experimental studies of surface wave breaking and air entrainment. In *Gas Transference at Water Surfaces*. W. Brutsaert and G. Jirka (Eds.), Reidel Press, Dordrecht, pp. 211–219.

Thorpe, S. A. (1982). "On the clouds of bubbles formed by breaking wind-wavesin deep water, and their role in air-sea gas transfer." Phil. Trans. Roy. Soc. Lond. A304: 155–210.

Thorpe, S. A. (1983). Bubble clouds: A review of their detection by sonar, of related models, and of how Kv may be determined. In *Oceanic Whitecaps and Their Role in Air-Sea Exchange Processes*. E. C. Monahan and G. MacNiocaill (Eds.), Reidel in association with Galway University Press, pp. 57–68, 1986.

Thorpe, S. A. (1986). "Measurements with an automatically recording inverted echo sounder: Aries and bubble clouds." J. Phys. Ocean. 16: 1462–1478.

Urick, R. J. (1984). Ambient Noise in the Sea, Undersea Warfare Technology Office, NAVSEA, D.O.N., Washington, DC (Also available from Peninsula Publishing, Los Altos, CA, 1986.).

Wenz, G. M. (1962). "Acoustic ambient noise in the ocean: Spectra and sources." J. Acoust. Soc. Am. 34: 1936–1956.

Wilson, J. H. (1983). "Wind-generated noise modeling." J. Acoust. Soc. Am. 73(1): 211–216.

Wu, J. (1980). "Wind-stress coefficient over sea surface near neutral conditions – a revisit." J. Phys. Oceano. 10: 727–740.

Wu, J. (1981). "Bubble populations and spectra in the near-surface ocean: Summary and review of field measurements." J. Geophys. Res. 86(C1): 457–463.

Wu, J. (1985). "Parameterization of wind-stress coefficients over water surfaces." J. Geophys. Res. 90(C5): 9069–9072.

Wu, J. (1986). Whitecaps, bubbles, and spray. In *Oceanic Whitecaps*. E. C. Monahan and G. MacNiocaill (Eds.), D. Reidel Publishing Co., Boston, MA, pp. 113–123.

Chapter 3
Fundamental Mechanisms

The discussion in the previous chapter indicates an exact treatment of this air–sea boundary interaction zone is not possible. If such a characterization were possible, one would still have the problem of environmental uncertainty associated with the measured boundary layer parameters required for comparison and evaluation. With this uncertainty, is it possible to examine qualitatively the possible mechanisms responsible for noise from this zone? This question is the motivation for this chapter, that is, to identify the physical mechanisms of sound production above the sea surface, below the sea surface, and by the motion of the surface itself. The approach provides a theoretical framework that enables the understanding of the multiple theories published on a variety of noise production mechanisms. The fundamentals of each particular approach and rationale can then be facilitated in terms of this general theoretical framework.

The Inhomogeneous Wave Equation and Sources of Sound

The description of the fundamental sources of sound requires the development of the inhomogeneous wave equations including sources. This chapter outlines the development of the time-dependent equation using indicial notation, identifies sources of sound, and develops analytical source integral expressions in the vicinity of the sea surface. These source integrals are shown to be of a monopole, dipole, and quadrupole nature. The complete derivation of these equations is presented in Appendix A. The derivation is classical, starting with conservation of mass.

The conservation of mass equation can be written on the basis of a balance of mass influx, mass efflux, and a source as shown in Fig. 3.1:

$$d\rho'/dt + \rho'\partial v_i/\partial x_i = \partial \rho'/\partial t + \partial \rho' v_i/\partial x_i = q. \tag{1}$$

This equation states the change in density, ρ', with time plus the net efflux of mass, $\partial \rho' v_i/\partial x_i$, equals the mass variation, q, within the incremental volume. Mass can not be created or destroyed in our treatment; however, a volume pulsation can result in a fluctuating mass or density within the volume.

W.M. Carey, R.B. Evans, *Ocean Ambient Noise*, Underwater Acoustics,
DOI 10.1007/978-1-4419-7832-5_3, © William M. Carey and Richard B. Evans 2011

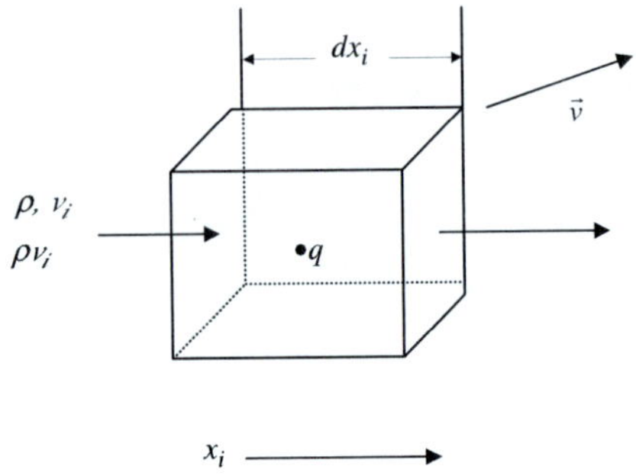

Fig. 3.1 The basis control volume geometry for the x_i direction

The second conservation equation is the conservation of momentum, and referring to Fig. 3.1, the rate of change of momentum in the control volume ($d\rho v'/dt + \partial\rho v'_j/\partial x_j$) is equal to the sum of the applied forces: gravitational forces ($g_i = -g\partial x_j\delta_{ij}/\partial x_j$), external forces, and the gradient of the stresses. Again

$$\begin{aligned} d\rho v'/dt + \partial\rho v'_j/\partial x_j &= \rho' g_i + f_{ei} - \partial p'_{ij}/\partial x_j; \\ \partial\rho' v'_i/\partial t + \partial\rho' v_i v_j/\partial x_j &= \rho' g_i + f_{ei} - \partial p'_{ij}/\partial x_j. \end{aligned} \tag{2}$$

The procedure is to perform a derivatization with respect to time of the equation of the conservation of mass, Eq. (1), a spatial derivatization of the equation of conservation of momentum, Eq. (2), and to subtract one from the other to yield

$$\partial^2\rho'/\partial t^2 = \partial q/\partial t - \partial f_{ei}/\partial x_i + \partial^2(\rho' v_i v_j + p'_{ij})/\partial x_i\partial x_j. \tag{3}$$

This equation does not include thermal, electromagnetic, and nonlinear factors.

The equation governing the acoustic quantities can be determined by assuming $\rho' = \rho_o + \rho$, $p' = P + p$, and $v_i = U_i + u_i$. The quantities ρ_o, P, and U are zero-order hydrodynamic quantities; whereas ρ, p, and u_i are first-order quantities much smaller than the zero-order ones. In addition, irrotational flow is assumed outside the source region.

$$\begin{aligned} &\vec{\nabla} \times \vec{v} = e_{ijk}\partial v_k/\partial x_j = 0, \\ &e_{123} = e_{231} = e_{312} = 1 \text{ and } e_{132} = e_{213} = e_{321} = -1 \end{aligned}. \tag{4}$$

Substitution of these quantities into the equation for ρ' and grouping terms of like order yields for the first-order quantities

$$\partial^2\rho/\partial t^2 = \partial q/\partial t - \partial f_{ei}/\partial x_i + \partial^2 T'_{ij}/\partial x_i\partial x_j. \tag{5}$$

This equation describes the first-order fluctuating quantities, and for a compressible fluid one may subtract $c^2\partial^2\rho/\partial x_i^2 = \partial^2 c^2\rho\delta_{ij}/\partial x_i\partial x_j$ from both sides of this equation to obtain the final inhomogeneous wave equation:

$$\partial^2\rho/\partial t^2 - c^2\partial^2\rho/\partial x_i^2 = \partial q/\partial t - \partial f_i/\partial x_i + \partial^2 T_{ij}/\partial x_i \partial x_j = 4\pi f(\tilde{x},t), \quad (6)$$

where $T_{ij} = 2\rho_o U_i u_j + \rho_o u_i u_j + (p - c^2\rho)\ \delta_{ij} - \mu[D_{ij} - 2/3\Theta\delta_{ij}]$ is the stress tensor and a simplified version is commonly referred to as the Lighthill stress tensor. [Eqs. (7) and (8) in Appendix A]. Finally, since the acoustic quantities assuming an isentropic process, $p = c^2\rho$, this equation is rewritten in terms of p

$$[\partial^2/\partial x_i^2 - (1/c^2)\partial^2/\partial t^2]p = -4\pi f(\tilde{x},t) \quad (7)$$

This inhomogeneous equation has on the right-hand side the source terms associated with the production of sound. The first term is the unsteady mass flow in the fluid and is of a monopole nature. The second term represents the divergence of a force on a boundary or that of an oscillating body, and is of a dipole nature. The third term represents the role of the viscous stresses and is the term Lighthill recognized as a quadrupole source.

These sources can be expressed mathematically as a point source $g = \exp(ikr)/r$ and its spatial derivatives. These higher-order sources of sound and their efficiencies are shown in Table 3.1. The problem of the noise production near, on, or above the sea surface requires knowledge of the relation between the actual sources of sound, the physical processes observed in the interaction zone, and these mathematical expressions. The required expressions for noise production will follow from the solutions of the inhomogeneous wave equation.

Table 3.1 Fundamental sources

Order	Term	Expression	Efficiency	Schematic	Description
0	Monopole	$\partial^o g/\partial x_{oi}^0$	$\eta_{rad} = (ka)$	 $r = a$	Volume fluctuation
1	Dipole	$\partial g/\partial x_{oi}$	$\eta_{rad} = (ka)^3$	O→O	Fluctuating force
2	Quadrupole	$\partial^2 g/\partial x_{oi}\partial x_{oj}$	$\eta_{rad} = (ka)^5$	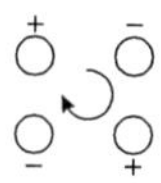	Turbulence distortion and rotation

The Point Source Solution, Green's Function

The inhomogeneous equation with a point source of sound is Green's equation:

$$[\partial^2/\partial x_i^2 - (1/c^2)\partial^2/\partial t_o^2]G(\tilde{x},t|\,\tilde{x}_o,t_o) = -4\pi\delta(\tilde{x}-\tilde{x}_o)\delta(t-t_o). \quad (8)$$

Green's function, $G(\tilde{x},t|\tilde{x}_o,t_o)$, describes the field at a point $\tilde{x},t$ from a point source at $\tilde{x}_o, t_0$. The function has two parts, $G(\tilde{x},t|\,\tilde{x}_o,t_o) = g(\tilde{x},t|\,\tilde{x}_o,t_o) + \chi(\tilde{x},t|\,\tilde{x}_o,t_o)$. The solution g is the free-space solution, no boundaries, whereas

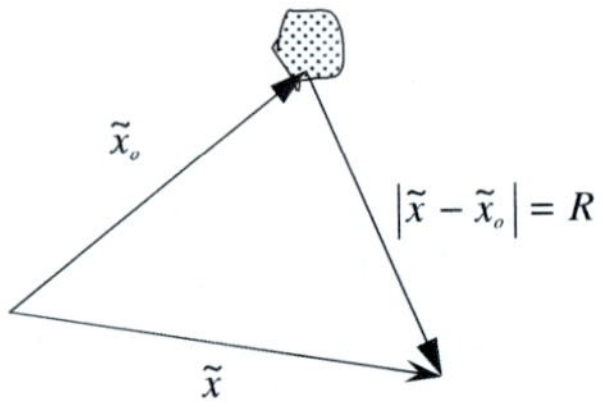

Fig. 3.2 The free-space geometry

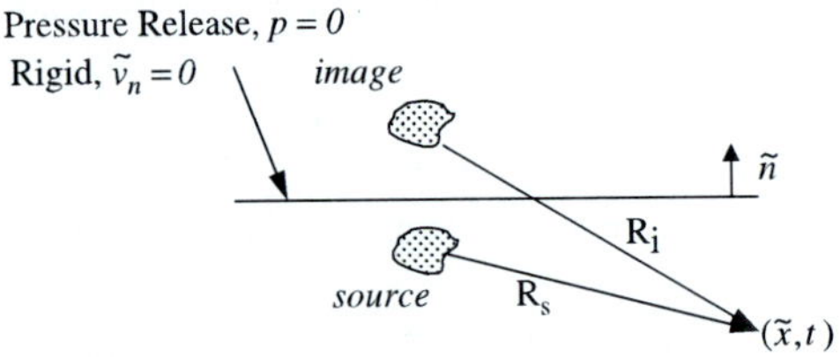

Fig. 3.3 The boundary value solution

the function χ is the boundary value solution to the homogeneous wave equation that includes surfaces. The free-space form of Green's function is, Fig. 3.2, $g(\tilde{x}, t|\tilde{x}_o, t_o) = \delta(t_o - (t - R/c))/R$. This solution describes the propagation of a sonic impulse emitted at $\tilde{x}_o, t_o$ and observed at time t at a point $\tilde{x}$ a distance R from the source. When a boundary is present, Fig. 3.3, the χ solution is required. The solutions are realized with the use of the method of images.

When the boundary condition is $G = 0$, a pressure-release boundary, the image has sign opposite that of the source and the solution is

$$G = g + \chi \; ; \; g = \delta(t_o - (t - R_s/c))/R_s \; , \; \chi = -\delta(t_o - (t - R_i/c))/R_i. \tag{9}$$

When the boundary condition is a rigid one, $\partial G/\partial n = 0$, the solution is

$$G = g + \chi \; ; \; g = \delta(t_o - (t - R_s/c))/R_s \; , \; \chi = +\delta(t_o - (t - R_i/c))/R_i. \tag{10}$$

The sea surface requires a mixed boundary condition. However, one may consider the noise sources in the air above the surface to reflect from a rigid boundary, whereas sources of sound beneath the surface reflect from a pressure-release surface.

Solutions to the Inhomogeneous Equation

A standard method of solving for the sound field radiated from a region with sources of sound is to multiply the pressure wave equation by Green's function and the Green's function equation by the pressure:

$$G[\partial^2/\partial x_i^2 - (1/c^2)\partial^2/\partial t^2]\,p = -4\,\pi\, f(\tilde{x}, t)G \tag{11}$$

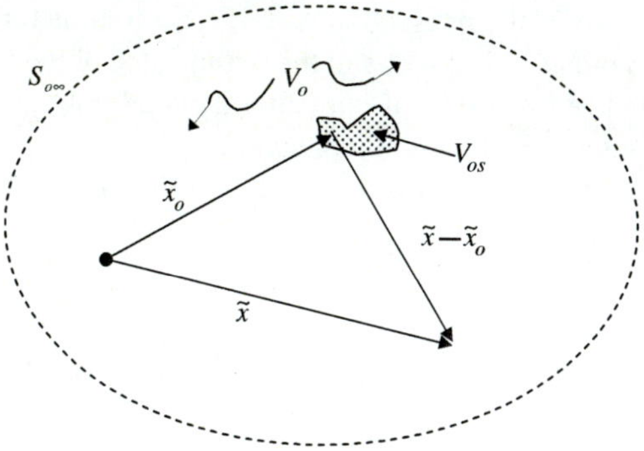

Fig. 3.4 The geometry

$$p[\partial^2/\partial x_i^2 - (1/c^2)\partial^2/\partial t^2]G(\tilde{x}, t|\tilde{x}_o, t_o) = -4\pi\delta(\tilde{x} - \tilde{x}_o)\delta(t - t_o)p. \quad (12)$$

Subtraction of the second from the first and interchanging variables yields

$$\begin{aligned} G[\partial^2/\partial x_{oi}^2 &- (1/c^2)\partial^2/\partial t_o^2]p - p[\partial^2/\partial x_{oi}^2 \\ &- (1/c^2)\partial^2/\partial t_o^2]G(\tilde{x}, t|\tilde{x}_o, t_o) = \\ &-4\pi f(\tilde{x}_o, t_o)G + 4\pi\delta(\tilde{x} - \tilde{x}_o)\delta(t - t_o)p. \end{aligned} \quad (13)$$

When no boundaries are present, Fig. 3.4, this equation can be integrated. Since $g(\tilde{x}, t|\tilde{x}_o, t_o) = \delta(t_o - (t - R/c))/R$, one can integrate over the interval of time, $0 \le t_o \le t + \varepsilon$. The choice of the upper limit $t + \varepsilon$ is to ensure that the delta function lies within the limits of integration. In addition, the sources are contained in a region within V_o defined as V_{os}, with $f(\tilde{x}_o, t_o) = 0$ outside this region.

Because of the properties of the delta function, the integral form of the above becomes

$$\begin{aligned} 4\pi p(\tilde{x}, t) = \int_o^{t+\varepsilon} dt_o \cdot \iiint_{V_o} dV_o[g[o]_o^2 p - p[o]_o^2 g] + 4\pi \int_o^{t+\varepsilon} dt_o \cdot \iiint_{V_o} dV_o f(x_o, t_o)g \\ \text{with} \quad [o]_o^2 \equiv [\partial^2/\partial x_{oi}^2 - (1/c^2)\partial^2/\partial t_o^2]. \end{aligned} \quad (14)$$

The first integral can be simplified by use of the divergence theorem:

$$\begin{aligned} \int_o^{t+\varepsilon} dt_o \cdot \iiint_{V_o} dV_o[g[o]_o^2 p - p[o]_o^2 g] = \int_o^{t+\varepsilon} dt_o \cdot \iint_{S_o} dS_o\{\hat{n}_o \cdot [g\tilde{\nabla}_o p - p\tilde{\nabla}_o g]\} \\ -(1/c^2) \iiint_{V_o} dV_o \cdot \int_o^{t+\varepsilon} dt_o(p\partial^2 g/\partial t_o^2 - g\partial^2 p/\partial t_o^2). \end{aligned} \quad (15)$$

Since the only boundary is the surface at a great distance $S_{o\infty}$, the first integral goes to zero because of the radiation condition. The second integral can be integrated as follows:

$$\begin{aligned} (1/c^2) \iiint_{V_o} dV_o \cdot \int_o^{t+\varepsilon} dt_o(p\partial^2 g/\partial t_o^2 - g\partial^2 p/\partial t_o^2) \\ = (1/c^2) \iiint_{V_o} dV_o \cdot [p\partial g/\partial t_o - g\partial p/\partial t_o]_o^{t+\varepsilon}. \end{aligned} \quad (16)$$

The limits on the result of the integration with respect to time represent the field at a time before the sound arrives, $t+\varepsilon$, and the initial value of the field at $t_o = 0$. The first limit must equal zero because of our causality conditions. The lower limit can be ignored if one assumes that these initial conditions existed in the distant past and are no longer important. The result is an expression for the radiated sound from the source region:

$$p(\tilde{x},t)=\iiint_{V_o} dV_o \cdot \int_o^{t+\varepsilon} dt_o f(x_o,t_o) g=\iiint_{V_o} dV_o \cdot \int_o^{t+\varepsilon} dt_o f(x_o,t_o)\delta(t_o-(t-R/c))/R. \tag{17}$$

Because the delta function is contained within the limits of integration, the result is the integrand evaluated at $t_o = t - r/c$, the retarded time, yielding

$$\int_o^{t+\varepsilon} dt_o f(x_o,t_o)\delta(t_o-(t-R/c))/R = f(x_o,t_o)/R|_{t_o=(t-R/c)} = [f(x_o,t_o)]_{ret}/R. \tag{18}$$

The radiated pressure field from the source region since f is zero outside V_{os} is

$$p(\tilde{x},t) = \iiint_{V_o} dV_o \cdot [f(x_o,t_o)]_{ret}/R = \iiint_{V_{os}} dV_{os} \cdot [f(x_o,t_o)]_{ret}/R. \tag{19}$$

Substitution for the source function $f(\tilde{x}_o,t_o)$ gives

$$\begin{aligned} P(\tilde{x},t) &= \textstyle\iiint_{V_o} dV_{os} \cdot [f(x_o,t_o)]_{ret}/R \\ &= (1/4\pi) \textstyle\iiint_{V_{os}} [\partial q/\partial t_o]_{ret} dV_{os}/R \\ &\quad -(1/4\pi) \textstyle\iiint_{V_{os}} [\partial f_{ei}/\partial x_{oi}]_{ret} dV_{os}/R \\ &\quad +(1/4\pi) \textstyle\iiint_{V_{os}} [\partial^2 T_{ij}/\partial x_{oi}\partial x_{oj}]_{ret} dV_{os}/R. \end{aligned} \tag{20}$$

This equation can be simplified by recognizing $\partial R/\partial x_{oi} = (x_i - x_{oi})/R = -\partial R/\partial x_i$. The final result in the absence of boundaries is dramatically simple:

$$\begin{aligned} 4\pi\, p(\tilde{x},t) = & \textstyle\iiint_{V_{os}} [\partial q/\partial t_o]_{ret} dV_{os}/R \quad \{A\ monopole\ term) \\ & -(\partial/\partial x_i) \textstyle\iiint_{V_{os}} [f_{ei}]_{ret} dV_{os}/R\ (A\ dipole\ term) \\ & +(\partial^2/\partial x_i \partial x_j) \textstyle\iiint_{V_{os}} [T_{ij}]_{ret} dV_{os}/R\ (A\ quadrupole\ term) \end{aligned} \tag{21}$$

The result has fundamental significance as it states a general principle of acoustics for compact sources. Since the order of the pole is equal to the order of the derivative of the point-source solution as shown in Table 3.1, each of these terms corresponds to a monopole, a dipole, and a quadrupole.

The Sea Surface

The previous solutions have not accounted for the presence of the sea surface and the method of images was proposed to account for the effect of the surface on the source radiation characteristics. The sea surface can be considered by using Green's function $G = g + \chi$. The result is an identification of the possible sources of sound and their basic characteristics.

The derivation of the surface integrals governing the production of sound is completely described in Appendix A. Here the derivation will simply be outlined for the sake of clarity in the discussion. The approach adopted is based on the traditional approach for the solution of the inhomogeneous wave equation as found in Stratton (1941) and Jackson (1962). In addition, the works of Curle (1955) and Huon-Li (1981) were relied on for the mathematical method and to ensure that these results were consistent with theirs. One major difference in these previous works is the use of both Laplace and Fourier transforms to derive the time-dependent form of Green's function and the use of the time-retarded functions to solve the inhomogeneous wave equation in the presence of the pressure-release surface. An alternative but comprehensive treatment of the problem may be found in the work of Cato (1991) that follows the Lighthill method coupled with Heaviside functions to handle the water–air interface and results in the description of the volume-distributed quadrupoles, surface dipoles, and distributed surface monopoles.

In the previous treatment in this chapter, the integral

$$\begin{aligned}(1/4\pi) &\int_o^{t+\varepsilon} dt_o \iiint_{V_o} dV_o[G[\circ]_o^2 P - P[\circ]_o^2 G] \\ &= (1/4\pi) \int_o^{t+\varepsilon} dt_o \iiint_{V_o} dV_o[G(\partial^2/\partial x_{oi}^2)P - P(\partial^2/\partial x_{oi}^2)G] \\ &\quad -(1/4\pi c) \int_o^{t+\varepsilon} dt_o \iiint_{V_o} dV_o[G(\partial^2/\partial t_o^2)P - P(\partial^2/\partial t_o^2)G]\end{aligned} \tag{22}$$

was equal to zero because of the initial conditions and because no surface was present. Again, the initial condition integral is equal to zero; however, the surface integral is not. Note that Green's function $G = g + \chi$ is being used.

The reason this initial condition integral is equal to zero is that the upper limit is greater than the interval of the argument of the delta function. The lower limit is the initial condition which we set to zero because time zero is so far in the distant past that any initial perturbations have died out. The final result is the statement

$$(1/4\pi c) \iiint_{V_o} dV_o[G(\partial/\partial t_o)P - P(\partial/\partial t_o)G]_o^{t+\varepsilon} = 0. \tag{23}$$

The first integral may be simplified by use of the divergence theorem to yield a surface integral containing the boundary conditions with l_{oi}, the ith directed normal:

$$\begin{aligned}&(1/4\pi) \int_o^{t+\varepsilon} dt_o \iiint_{V_o} dV_o[G(\partial^2/\partial x_{oi}^2)p - p(\partial^2/\partial x_{oi}^2)G] \\ &\rightarrow (1/4\pi) \cdot \iint_{S_o} dS_o\{(l_{oi}/r)[\partial p/\partial x_{oi}]_{g,\chi} + (l_{oi}/r^2)(\partial r/\partial x_{oi})[p]_{g,\chi} \\ &\qquad\qquad +(l_{oi}/rc)(\partial r/\partial x_{oi})[\partial p/\partial t_o]_{g,\chi}\}\end{aligned} \tag{24}$$

where $[f]_{g,\chi} = \int_o^{t+\varepsilon} [f\delta(t_o - (t - r_s/c) + f\delta(t_o - (t - r_i/c))]dt_o$ defines the retardation. The $(1/r)$ factor has been treated separately in this analysis and the assumption $|r_s| = |r_i|$ has been made, that is, the source is close to the sea surface, $\{(x_{o3})_s/\lambda \leq 1\}$. This integral can be further simplified as

$$\rightarrow (1/4\pi) \iint_{S_o} dS_o\{(l_{oi}/r)[\partial p\delta_{ij}/\partial x_{oj}]_{g,\chi} + (1/4\pi)\cdot(\partial/\partial x_i) \iint_{S_o} dS_o l_{oj}[p\delta_{ij}]/r \tag{25}$$

The volume integral can be treated as before except for the change in Green's function:

$$\int_o^{t+\varepsilon} dt_o \iiint_{V_o} dV_o F(\tilde{x}_o, t_o) G(\tilde{x}_o, t_o\,|\tilde{x}, t) = \iiint_{V_o} dV_o [F(\tilde{x}_o, t_o)]_{g,\chi}/r. \tag{26}$$

The function $F(\tilde{x}_o, t_o)$ when substituted yields three integrals corresponding to the monopole, dipole, and quadrupole terms that compose this function (see Appendix A for the derivative identities to obtain these relationships):

$$4\pi\ F(\tilde{x}_o, t_o) = \partial q/\partial t_o - \partial f_{ei}/\partial x_{oi} + \partial^2 T_{ij}/\partial x_{oi}\partial x_{oj}. \tag{27}$$

The first term is already in its simplest form; the second and third terms can be manipulated to yield the following surface integrals:

(a) The external force term,

$$\begin{aligned}&(-1/4\pi) \iiint_{V_o} dV_o (1/r)[\partial f_{ei}(\tilde{x}_o, \tau)/\partial x_{oi}]_{g,\chi}\cdot\\ &= (-1/4\pi) \iint_{S_o} dS_o([f_{ei}]_{g,\chi}/r) l_{oi} + (1/4\pi)\partial/\partial x_i \iiint_{V_o} ([f_{ei}]_{g,\chi}/r) dV_o;\end{aligned} \tag{28}$$

(b) The stress tensor terms,

$$\begin{aligned}&(1/4\pi) \iiint_{V_o} dV_o/r\cdot[\partial^2 T_{ij}/\partial x_{oi}\partial x_{oj}]_{g,\chi} = (1/4\pi)(\partial^2/\partial x_i \partial x_j) \iiint dV_o[T_{ij}]/r\\ &+(1/4\pi)(\partial/\partial x_i) \iint dS_o l_{oj}[T_{ij}]/r + (1/4\pi) \iint_{S_o} dS_o(l_{oi}/r)\partial[T_{ij}]/\partial x_{oj}.\end{aligned} \tag{29}$$

The Source Integrals

Collecting like terms, the final result is that the radiated acoustic field is equal to the sum of five integrals:

$$4\pi \cdot P(\tilde{x}, t) = \sum_{q=1}^{5} I_q(\tilde{x}, t). \tag{30}$$

$I_1(\tilde{x}, t)$: The first integral, $I_1(\tilde{x}, t)$, represents the fluctuations of mass caused by the volume pulsation of a bubble or a bubble cloud and this basic source is a monopole. Because of the surface image effect carried in the $[\circ]_{g,\chi}$ operator, this term has a dipole characteristic that is referred to here as a doublet.

$$I_1(\tilde{x}, t) = \iiint_{V_o} dV_o/r \cdot [\partial q/\partial t_o]_{g,\chi} \tag{31}$$

Bubbles are produced by breaking waves, spray, splash, and rain. Experimental evidence indicates that it is the initial response of the bubble resulting from its entrainment near the sea surface that is responsible for the radiation of sound. Likewise, microbubble clouds are formed by spilling and plunging breakers. The properties of these clouds are described in Chapter 2 (see Figs. 2.5 and 2.6) and Appendix D. They are simply formed and entrained in the subsurface orbital motion of the gravity wave. It is the initial entrainment and compression that is responsible for the radiation of sound. These two mechanisms describe the production of noise at frequencies below 200 Hz, individual bubbles at frequencies above 500 Hz, and both mechanisms in the intermediate-frequency range.

The transient response of a bubble, $a_1(t)$, has been treated by Fitzpatrick and Strasberg (1956), and Strasberg (1956) (also see Appendix E) and is known to be a damped sinusoidal motion with a resonant frequency Minnaert (1933):

$$\begin{aligned} a_1(t) &= A\, exp(-\zeta\, t/2)\, sin(\omega_d t + \psi) \\ A &= [a_1(o)^2 \omega_o^2/\omega_d^2 + \dot{a}_1(o)^2/\omega_d^2 + \dot{a}_1(o) a_1(o)\zeta/\omega_d]^{1/2}; \end{aligned} \tag{32}$$

$$\zeta = r/m_e, \;\; \omega_o^2 = k_e/m_e = 3\gamma\, p_o/a_o^2\rho, \quad \omega_d = \omega_o\sqrt{1 - (\zeta/2\omega_o)^2}. \tag{33}$$

Bubble clouds that are acoustically compact have an analogous response with a resonance determined by a modified Minnaert frequency, $\omega_o^2 = 3\gamma\, p_o/a_o^2\rho\, \chi(1-\chi)$, where χ is the volume fraction, described in Appendix D and Carey (1993). The image interference effect due to the sea surface is described in Appendix G.

$I_2(\tilde{x}, t)$: The next integral, $I_2(\tilde{x}, t)$, represents the role an external force can have when acting on the volume:

$$I_2(\tilde{x}, t) = -(\partial/\partial x_i) \iiint_{V_o} ([f_{ei}]_{g,\chi}/r) dV_o. \tag{34}$$

By itself, one recognizes that it has a dipole characteristic. However, this dipole characteristic is increased in order by the presence of the sea surface contained in the $[\circ]_{g,\chi}$ operator. Thus, this term may not be important at large distances from the interaction zone.

$I_3(\tilde{x}, t)$: The integral $I_3(\tilde{x}, t)$ is basically a quadrupole source term that with its reflection in the sea surface becomes higher order and is thus not important in the far field of the interaction zone.

$$I_3(\tilde{x},t) = (\partial^2/\partial x_i \partial x_j) \iiint dV_o([T_{ij}]_{g,\chi}/r) \tag{35}$$

Turbulent flows containing microbubbles are known to produce sound. The microbubbles provide the compressibility and the turbulence provides the mechanical energy. This mechanism was discussed by Carey (1988), based on a treatment of turbulent flow with microbubbles by Ffowcs Williams (1969), and was shown not to be important because of the proximity of this type of turbulent flow to the sea surface. When there is no pressure-release surface present, this term can represent the pressure fluctuations of the quadrupole field due to oceanic turbulence and at the low frequency may place a lower bound on the noise measured with a hydrophone.

The air–sea interaction zone includes the turbulent boundary layer fluctuations in the region above and adjacent to the sea surface. Ffowcs Williams and Guo (1988) applied this integral to the turbulent airflow above the sea surface. In this case $T_{ij} \approx \rho_a u_i u_j$ and $p_{ij} - c_a^2 \rho \delta_{ij} \approx 0$, the $I_3(\tilde{x}, t)$ becomes

$$\begin{aligned} I_3(\tilde{x},t) &= (\partial^2/\partial x_i \partial x_j) \textstyle\iiint dV_o[T_{ij}]_{g,\phi}/r = (\partial^2/\partial x_i \partial x_j) \textstyle\iiint dV_o[\rho_a u_i u_j]_{g,\phi}/r \\ &\textstyle\iiint dV_o(\partial^2/\partial x_i \partial x_j\{[\rho_a u_i u_j]_{g,\phi}/r\} \\ &\rightarrow (x_i x_j \rho_a/r^3 c_a^2)(\partial^2/\partial t^2) \textstyle\iiint dV_o[u_i u_j]_{g,\chi}. \end{aligned} \tag{36}$$

This can be used to calculate the pressure fluctuations on the sea surface resulting from the air turbulence in the volume V_o using the plane wave transmission coefficient

$$4\pi p(\tilde{x},t) = (2\rho_w c_w/(\rho_a c_a + \rho_w c_w))(x_i x_j \rho_a/r^3 c_a^2)\{\partial^2/\partial t^2 \iiint dV_o[u_i u_j]_{g,\chi}\}. \tag{37}$$

Ffowcs Williams stated that if the flow velocity is u and the length scale of a turbulent eddy is l, then $\partial/\partial t$, the frequency scale, should be on the order of u/l and

$$4\pi p(\tilde{x},t) \propto (\rho_a/c_a^2)(lM^4/\,|x|). \tag{38}$$

Since the turbulent boundary layer has a very small Mach number, M, and since the difference between the impedance of water and air is large, very little of the atmospheric boundary layer pressure fluctuation is coupled to the water. Kuryanov (1990) presented an analytical development using the space-time correlative properties for flow over a rigid surface to estimate the frequency wavenumber spectral density. The integral of this wavenumber spectral density restricted to a small wavenumber region, $|\tilde{\kappa}| < \omega/c$, determined by a radiation condition yields the spectral density and when coupled with a high-frequency surface wave spectrum produces a result that the radiated acoustic spectral density is proportional to $M^{4.5}$. When applied to frequencies less than 100 Hz, noise levels were expected to be about 50 $dB\ re\ (\mu Pa)^2/Hz$ with no frequency dependence. Yen and Perrone (1979) combined the integral presented here with a space-time correlation function based on the use of the Kolmogroff hypothesis, isotropic turbulence, and a boundary layer model to estimate the wind speed profile and the thickness of the turbulent boundary layer to obtain the wavenumber spectrum proportional to $\omega^{4.5}$and an estimated level of $55\,dB\,re\,(\mu P_a)^2/Hz\ at\ f = 10\,Hz$. The turbulent boundary layer radiated sound is expected to be a weak contributor to the ambient noise field above 10 Hz, but can be a contributor at frequencies less than 5 Hz.

$\boldsymbol{I_4(\tilde{x}, t)}$: The integral $I_4(\tilde{x}, t)$ represents the rate of change of momentum at the sea surface. Because the image and source velocities have opposite signs, this integral behaves as a dipole in the surface:

$$I_4(\tilde{x}, t) = -\iint dS_o(l_{oi}/r)[\partial \rho u_i/\partial t_o]_{g,\chi}. \tag{39}$$

Guo and Ffowcs (1990, Eq. (1)) used the Kirchhoff theorem (which provides a result equivalent to the one given above) to formulate the radiated pressure from a drop impact:

$$4\pi \cdot \cdot p(\tilde{x}, t) = 2\rho_o(\partial/\partial t)\iint u_3(\tilde{x}_o, \tau)dx_{o1}dx_{o2}/r;\ \ r = |\tilde{x}_o - \tilde{x}|_{x_3=o};\ \tau = t - r/c. \tag{40}$$

This integral has been shown to describe the radiated sound from a drop impact and thus provides an analytical description of spray, splash, and rain noise. The role of impacts in the higher-frequency ambient noise spectrum is considered significant.

$\boldsymbol{I_5(\tilde{x}, t)}$: Finally, the integral $I_5(\tilde{x}, t)$ has a dipole characteristic indicated by the $(\partial/\partial x_i)$term:

$$I_5 = (\partial/\partial x_i) \cdot \iint dS_o(l_{oi}/r)[2\rho_o U_i u_j + \rho_o u_i u_j + p\delta_{ij}]_{g,\chi}. \tag{41}$$

Because the viscous terms of the stress tensor have been ignored, the pressure term is identically zero. The flow velocities retain directional information such that $[2\rho_o U_i u_j + \rho_o u_i u_j]_{g,\chi} \neq 0$. These terms represent the wave–turbulence and turbulence interaction. The turbulence interaction is second order and can be neglected, whereas the wave–turbulence interaction may have a role at infrasonic frequencies. However, in the derivation of this expression, the quantity $\rho_o U_i U_j$ was treated as a zero-order quantity. This term describes the wave–wave interaction and is much larger than the wave–turbulence term; when it is reintroduced

$$[2\rho_o U_i u_j + \rho_o u_i u_j]_{g,\chi} \to [2\rho_o U_i u_j + \rho_o U_i U_j]_{g,\chi}.$$

Yen and Perrone (1979) examined both of these mechanisms and concluded that they can only be relevant at frequencies less than 10 Hz. The wave–turbulence contribution can exceed the wave–wave contribution at frequencies less than 5 Hz, but is still expected to be less than the wave–wave interaction. At 10 Hz, the noise level from the wave–turbulence interaction was estimated to be about $70\,dB\,re\,(\mu\,P_a)^2/Hz$. Needless to say, these estimates are very uncertain, largely owing to lack of knowledge of the oceanic turbulence characteristics in the air–sea interaction zone. Goncharov (1970) started with the first term in $I_5(\tilde{x},t)$ and for the wave (ζ_s)–turbulence (ζ_T) interaction obtained

$$\begin{aligned} 4\pi p(\tilde{x},t) &= (-2\rho)(\partial^3/\partial x_3\partial t^3)\int_{-\infty}^{\infty} dx_{o1}\int_{-\infty}^{\infty} dx_{o2}[\zeta_s][\zeta_T]/r; \quad r = |\tilde{x}-\tilde{x}_o|_{(x_{o3}=0)}; \\ \textit{wave displacement} \quad [\zeta]_s &= \zeta_s(x_{o1},x_{o2},t-r/c); \\ \textit{turbulence displacement} \quad [\zeta]_T &= \zeta_t(x_{o1},x_{o2},t-r/c). \end{aligned} \tag{42}$$

Goncharov then used the Pierson–Moskowitz (1964) wave spectrum, the Kolmogorov similarity hypothesis, and many algebraic manipulations to obtain the pressure spectrum $P(f) \approx (10^6/f^2)\,[\mu P_a^2/Hz]^{1/2}$. The noise level is about $80\,dB\,re\,(\mu P_a)^2/Hz$ at $10\,Hz$ and about $40\,dB\,re\,(\mu P_a)^2/Hz$ at $100\,Hz$; different from the Yen and Perrone (1979) result by *10 dB*.

The problem with these estimates is the knowledge of the oceanic turbulence spectral and correlative characteristics. The analysis discussed above shows that wave–turbulence interaction radiated noise is equivalent to the lower limit on the Wenz curve between 10 and 100 Hz.

Goncharov further showed that for the wave–wave interaction the following result can be obtained from I_5 when the zero-order surface waves are retained:

$$\begin{aligned} I_5 &= (\partial/\partial x_i)\cdot\iint dS_o(l_{oi}/r)[\rho_o U_i U_j]_{g,\chi} \\ &\to -\rho\,(\partial^3/\partial x_3\partial t^2)\cdot\int_{-\infty}^{\infty} dx_{o1}\int_{-\infty}^{\infty} dx_{o2}[\zeta_s]^2/r; \; x_{o3}=0. \end{aligned} \tag{43}$$

This expression is consistent with the expressions given by Longuet-Higgins (1950) and Brekhovskikh (1967). This mechanism has been thoroughly treated by Hughes (1976), Kibblewhite and Ewans (1985), and Cato (1991). Comparisons with experimental results will follow in Chapter 4.

References and Suggested Readings

Blokhintz, D. (1956). The Acoustics of an Inhomogeneous Moving Medium, NACA Technical Memorandum No. 1399.

Brekhovskikh, L. M. (1967). "Generation of sound waves in a liquid by surface waves." Sov. Phys. Acoust. 12(3): 323–350.

Carey, W. M. and J. W. Fitzgerald (1990). Low frequency noise from breaking waves. In *Natural Physical Sources of Underwater Sound.* Cambridge, UK. B. R. Kerman (Ed.), Kluwer Academic Publishers, Dordrecht (1993), pp. 277–304.

Carey, W. M. and J. W. Fitzgerald (1993). "Measurement of the sound produced by a tipping trough with fresh and salt water" J. Acoust. Soc. Am. 93(6): 3178–3192 (Also see Erratum, J. Acoust. Soc. Am. 94(5): 3018, 1993).

Cato, D. H. (1976). "Ambient sea noise in waters near Australia." J. Acoust. Soc. Am. 60(2): 320–328.

Cato, D. H. (1991a). "Sound generation in the vicinity of the sea surface: Source mechanisms and coupling to the received sound field." J. Acoust. Soc. Am. 89(3): 1076–1095.

Cato, D. H. (1991b). "Theoretical and measured underwater noise from surface orbital wave motion." J. Acoust. Soc. Am. 89(3): 1096–1112.

Curle, N. (1955). "The influence of solid boundaries upon aerodynamic sound." Proc. R. Soc. A. 231(1187): 505–514.

Dowling, A. P. and J. E. Ffowcs Williams (1983). *Sound and Sources of Sound*. Ellis Horwood Limited, Halsted Press-John Wiley & Sons, New York, NY, pp. 37–62, 146–166.

Ffowcs Williams, J. E. (1969). Hydrodynamic noise. In *Annual Review of Fluid Mechanics*. W. R. Sears and M. Van Dyke (Eds.), Annual Reviews, Inc. Palo Alto, CA, Vol. 1, pp. 197–222.

Fitzpatrick, H. M. and M. Strasberg (1956). Hydrodynamic sources of sound. In *Naval Hydrodynamics*. F. S. Sherman (Ed.), National Academy of Sciences-National Research Council Pub 515, LC-57-60045, pp. 241–280.

Franz, G. J. (1959). "Splashes as sources of sound in liquids." J. Acoust. Soc. Am. 31(8): 1080–1096.

Furduev, A. V. (1966). "Undersurface cavitation as a source of noise in the ocean." Atmos. Oceanic Phys. 2(235): 314–320.

Goncharov, V. V. (1970). "Sound generation in the ocean by the interaction of surface waves and turbulence." Izv., Atmos. Oceanic Phys. 6(11): 1189–1196.

Guo, Y. P. and J. E. Ffowcs Williams (1991). "A theoretical study of drop impact sound and rain noise." J. Fluid Mech. 227: 345–355.

Hinze, J. O. (1959). *Turbulence*. McGraw-Hill Book Co., New York, NY, pp. 13–25.

Hughes, B. (1976). "Estimates of underwater sound (and infrasound) produced by nonlinearly interacting ocean waves." J. Acoust. Soc. Am. 60(5): 1032–1039.

Hunt, F. V. (1955). "Notes on the exact equations governing the propagation of sound in fluids." J. Acoust. Soc. Am. 27(6): 1019–1039.

Huon-Li (1981). On Wind-Induced Underwater Ambient Noise, NORDA TN 89, NORDA, NSTL, MS.

Isakovich, M. A. and B. F. Kur'yanov (1970). "Theory of low frequency noise in the ocean." Sov. Phys. Acoust. 16(1): 49–58.

Jackson, J. D. (1962). *Classical Electrodynamics*. John Wiley & Sons, Inc., New York, NY, pp. 183–189.

Kibblewhite, A. C. and K. C. Ewans (1985). "Wave-wave interactions, microseisms, and infrasonic ambient noise in the ocean." J. Acoust. Soc. Am. 78(3): 981–994.

Kuryanov, B. F. (1990). *The Theory of Low-Frequency Noise Generated by Turbulence near the Atmosphere-ocean Interface*. Natural Physical Sources of Underwater Sound, Sea Surface Sound (2), Cambridge, UK. Kluwer Academic Publishers (1993): pp. 255–262.

Lighthill, M. J. (1952). "On sound generated aerodynamically, I: General theory." Proc. Roy. Soc. (Lond.) A211: 564–587.

Lighthill, M. J. (1979). *Waves in Fluids*. Cambridge University Press, Cambridge, UK.

Lloyd, S. P. (1981). "Underwater sound from surface waves according to the Lighthill-Ribner Theory." J. Acoust. Soc. Am. 69(2): 425–435.

Longuet-Higgins, M. S. (1950). "A theory of the origins of microseisms." Trans. Roy. Soc. A 243: 1–35.

Minnaert, M. (1933). "Musical air-bubbles and sounds of running water." Phil. Mag. XVI (7th Series): 235–249.

Morse, P. M. and K. U. Ingard (1961). Linear acoustic theory. In *Handbuch der Physik, Band XI/1, Akustik I*. E. Flügge (Ed.), Springer-Verlag, Berlin, pp. 1–128.

Morse, P. M. and K. U. Ingard (1968). *Theoretical Acoustics*. McGraw-Hill Book Company, New York, NY.

Nichols, R. H. (2005). "Some notable noise: Monsters and machines (Originally published in the office of naval research proceedings of the international workshop on low frequency propagation and noise, 1974)." IEEE J. Ocean. Eng. 30(2): 257–261.

Pierce, A. D. (1981). *Acoustics: An Introduction to its Physical Principles and Applications*. Acoustical Society of America, Woodbury, NY.

Prosperetti, A. (1988a). Bubble dynamics in oceanic noise. In *Sea Surface Sound*. B. R. Kerman (Ed.), Kluwer Press, Boston, pp. 151–172.

Prosperetti, A. (1988b). "Bubble-related ambient noise in the ocean." J. Acoust. Soc. Am. 84(3): 1042–1054.

Rayleigh, J. W. S. (1945). *Theory of Sound*. Dover Publications, New York, NY.

Reschevkin, S. N. (1963). *The Theory of Sound*. Pergamon Press, New York, NY.

Ross, D. H. (1976). *Mechanics of Underwater Noise*. Peninsula Publishing, Los Altos, CA.

Ross, D. H. (2005). "Ship sources of ambient noise (Originally published in the office of naval research proceedings of the international workshop on low frequency propagation and noise, 1974)." IEEE J. Ocean. Eng. 30(2): 257–261.

Strasberg, M. (1956). "Gas bubbles as sources of sound" J. Acoust. Soc. Am. 28(1): 20–26.

Stratton, J. D. (1941). *Electromagnetic Theory*. McGraw-Hill Book Company, Inc., New York, NY, pp. 424–429.

Yen, N. and A. J. Perrone (1979). Mechanisms an Modeling of Wind-Induced Low-Frequency Ambient Sea Noise, Naval Underwater Systems Center, New London, CT, NUSC Rept. TR 5873 (DTIC A-650-70-6502).

Chapter 4
The Measurement of Oceanic Ambient Noise

Introduction

Ambient noise investigations constitute one of the largest sections of the Journal of the Acoustical Society of America. Urick (1984) has summarized a good many of these experimental papers and his report is valuable as it updates and extends the work of Wenz (1972). In this chapter the discussion focuses on the key aspects of ambient noise by interpreting the experimental observations in light of the fundamental production mechanisms of ambient noise. The theoretical treatments of these source mechanisms can be found in Chapter 3, with the appendices containing detailed derivations. The source mechanisms are used in this chapter as part of our overview of the characteristics of measured ambient noise. Since the measurements span some 60 years, the following question arises: What was measured and how do these measurements compare with those currently performed?

Repeatable and accurate measurement of the noise background requires high-quality calibrated receivers. Calibrated hydrophones were not available prior to World War II; consequently, the ambient noise measurements discussed in this treatment were made during and after that war. In his history of underwater acoustics, Goodman (2004) points out that although magnetostriction and piezoelectricity were discovered in the nineteenth century, additional hydrophone technology development was required to convert acoustic pressures to amplified voltages that enabled the practical application of these electromechanical phenomena. The conductive and corrosive nature of sea water as well as the requirement to operate over a wide range of depths and pressures were additional technological difficulties to be overcome. Langevin, Chilowsky, and Fessenden produced transmitters capable of echo-ranging to a distance of miles. Fessenden used a moving coil transmitter. Langevin developed a resonant quartz receiver that was replaced by the unstable Rochelle salt and ammonium dihydrogen phosphate. Between 1940 and 1950, barium titanate piezoelectric ceramics replaced these in hydrophone applications. By the mid-1950s, two types of receivers were used: (1) moving-coil, oil-filled, omnidirectional hydrophones, and (2) the barium titanate hydrophones with a relatively flat response to the kilohertz region. The availability of hydrophones capable of operating in the ocean environment, even with primitive data acquisition and analysis systems, provided the technological tools necessary for oceanic ambient

W.M. Carey, R.B. Evans, *Ocean Ambient Noise*, Underwater Acoustics,
DOI 10.1007/978-1-4419-7832-5_4,

noise research. The further development of lead zirconate titanate enabled the development of wide-band sensitive hydrophones, which are widely used by investigators in the study of oceanic noise.

The first section of this chapter provides an overview of a measurement system that spans the evolution of technology to provide the basis for interpreting and comparing measured results from the war years to the present. The treatment is general but specific to the determination of the mean square pressure, the pressure spectral density, the correlation function, and other noise characteristics. The current acoustic definitions and conventions used in this chapter along with these quantities are found in Appendix B, where the American National Standards are discussed and applied. The second portion of this chapter is a phenomenological overview of ambient noise from the characteristic of Wenz (1962) to the present using selected references of noise studies.

Noise Measurement

The measurement of ambient noise has progressed from analog measurement to digitization and digital fast Fourier transform (FFT) analyses. Early measurements focused on the root mean square pressure, p_{rms}, mean square pressure $< p^2 >$, or the autocorrelation function, $\Gamma_p(\tau)$, in the specified band. A representative measurement schematic is shown in Fig. 4.1. The calibrated hydrophone converts the pressure fluctuations in the water to charge fluctuations, a charge-sensitive amplifier converts the signal to a time-varying voltage, and this temporal voltage variation is band-pass-filtered. The implementation of the filter was often the application scheme whereby the voltage was multiplied by $\cos(2\pi f_0 t)$ and $\sin(2\pi f_0 t)$. Since these quantities are orthogonal to one another at any time *t*, the cross-products integrate to zero and for all practical purposes can be treated as independent variables, with two degrees of freedom. The square of the voltage $v_o^2(t)$ is integrated over a time interval, T, for which conditions were considered stationary as determined by the correlative properties and the state of the sea. The output of this integration is then sampled and summed to provide an estimate of the mean square voltage $\langle v_o^2 \rangle$.

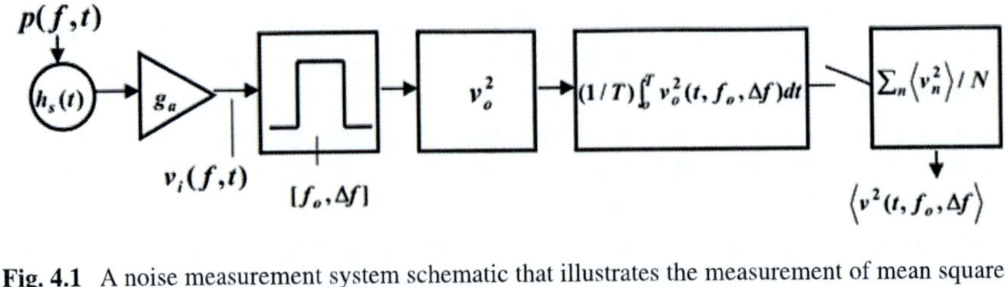

Fig. 4.1 A noise measurement system schematic that illustrates the measurement of mean square ambient noise pressures. This schematic can be applied to both analog and fast Fourier transform systems

It is readily seen by analogy that the estimate of the mean square voltage or mean square pressure in this single band Δf wide is equivalent to the complex product of FFTs, $P(f) \cdot P(f)^*$, of the pressure signal. However, sound that is continuous-bounded, nonperiodic, and stationary poses a problem for Fourier analysis. Fourier integrals are infinite integrals and consequently convergence must be considered. The continuous version of Parseval's theorem, Plancherel's theorem (see Middleton (1987, p 137), Appendix B),

$$\int_{-\infty}^{+\infty} |p(t)|^2 dt = 2\pi \cdot \int_{-\infty}^{+\infty} P(\omega) \cdot P(\omega)^* d\omega = \int_{-\infty}^{+\infty} P(f) \cdot P(f)^* df, \quad (1)$$

states the problem. If the integral of $|p(t)|^2$ converges, then the integral of $|P(f)|^2$ converges. If the pressure variation is random but stationary, $|p(t)|^2$does not diminish as $t \to \infty$ and the integral will not converge. However, if we restrict p(t) such that

$$\begin{array}{ll} p(t) = 0, & -\infty < t < -T/2 \;\; ; T/2 < t < +\infty \\ p(t) \neq 0, & -T/2 < t < +T/2 \end{array} , \quad (2)$$

then for large but finite T

$$\int_{-\infty}^{+\infty} |p(t)|^2 dt \to \int_{-T/2}^{+T/2} |p(t)|^2 dt = T{<}p(t)^2{>}_T = 2\pi \int_{-\infty}^{+\infty} |P(\omega)|^2 d\omega$$
$$= \int_{-\infty}^{+\infty} |P(f)|^2 df \quad (3)$$

where $P(f) = \int_{-T/2}^{+T/2} p(t) exp(-i2\pi f\, t)dt.$

These integrals decrease at a sufficient rate with $\omega = 2\pi f \to \infty$ to ensure convergence for large but not infinite values of T:

$$< p(t)^2 >_T = [2\pi/T] \int_{-\infty}^{+\infty} |P(\omega)|^2 / d\omega = [1/T] \int_{-\infty}^{+\infty} |P(f)|^2 df$$
$$< p(t)^2 >_T / 2\pi = \int_{-\infty}^{+\infty} (|P(\omega)|^2/T) d\omega \;\; and \;\; < p(t)^2 >_T = \int_{-\infty}^{+\infty} (|P(f)|^2/T) df$$
$$(|P(\omega)|^2/T) \;\; and \;\; (|P(f)|^2/T) \;\; are\ spectral\ densities\ per\ unit\ time. \quad (4)$$

For this case of bounded, nonperiodic, and stationary pressure fluctuations, one can define time-varying means and mean square quantities for large T:

$$< p(t) >_T = (1/T) \int_{-T/2}^{+T/2} p(t) dt \;\; and \;\; < p(t)^2 >_T = \int_{-\infty}^{+\infty} (|P(f)|^2/T) df. \quad (5)$$

The Autocorrelation Function, $\Gamma_p(\tau)$

The variation with time can be described by use of the autocorrelation function:

$$\Gamma_p(\tau) \equiv \lim_{T\to\infty} \left\lfloor (1/T) \int_{-T/2}^{+T/2} p(t)p(t+\tau)dt \right\rfloor; \\ \to \Gamma_p(0) \equiv \lim_{T\to\infty} \left[(1/T) \int_{-T/2}^{+T/2} p(t)^2 dt \right] = < p(t)^2 >_T . \tag{6}$$

The ensemble-averaged mean square value of $< p(t)>_T$ can be determined by the following product:

$$\ll p(t) >_T^2 >_e =< (1/T) \int_{-T/2}^{T/2} p(t_1)dt_1 \, (1/T) \int_{-T/2}^{T/2} p(t_2)dt_2 >_e . \tag{7}$$

The autocorrelation function can be related to this quantity:

$$\ll p(t) >_T^2 >_e = (1/T^2) \int_{-T/2}^{T/2} \cdot \int_{-T/2}^{T/2} \Gamma_p(t_2 - t_1) dt_2 dt_1 \to (2/T) \int_0^T (1-\tau/T)\Gamma_p(\tau)d\tau . \tag{8}$$

The Fourier transform of the autocorrelation function is $2\pi/T$ times the spectral density $|P(\omega)|^2$ of p(t):

$$F\left\{\Gamma_p(\tau)\right\} = (1/2\pi) \int_{-\infty}^{+\infty} \Gamma_p(\tau) exp(-i\omega\,\tau)d\tau \to 2\pi P(\omega)P(\omega)^*/T . \tag{9}$$

The Power Spectral Density

Continuing with measurement of stationary but random pressure fluctuations, one needs to consider sampling. Each observed time series of p(t) is one member, a sample member of the family of all possibilities, of the ensemble. Let the ensemble be represented by {p(t)} and let $p^j(t)$ be the jth sample of the random process {p(t)}. If the variations of the mean, mean square, and autocorrelation of p(t) exhibit significant variation with time, the process is nonstationary, if there are no significant variations with time, the process is weakly stationary, and if all moments of p(t) show no variation with time, the process is strongly stationary or stationary in the strict sense. If a member of a stationary process is representative of the whole ensemble, then the process can be considered ergodic.

For $p^j(t)$, which vanishes everywhere outside the interval $t_1 - T/2 < t < t_1 + T/2$, the average power, $< \Pi(T) >$, or average energy, $< E(T) >$, has no dependence on t over the interval and we have

$$\Pi^j(T) \equiv E^j(T)/T = (1/T)\int_{t_1-T/2}^{t_1+T/2} p^j(t)^2 dt. \tag{10}$$

According to Parseval's theorem,

$$\Pi^j(T) \equiv E^j(T)/T = \int_{-\infty}^{+\infty} \{|P^j(f)|^2/T\}df. \tag{11}$$

$$W^j_p(f)_T \equiv 2|P^j(f)_T|^2/T \rightarrow \Pi^j(T) = \int_o^{\infty} W^j_p(f)_T df. \tag{12}$$

Since the power spectral density, $W^j_p(f)_T$, is an even function of f, one does not need the negative frequencies and the factor of 2 in the definition above accounts for the change in the integration limits.
One can then define a linear average power as

$$<\Pi(T)>_{NT} = 1/N\sum_{j=1}^{N}\Pi^j(T) = 1/N\sum_{j=1}^{N}\int_o^{\infty} W^j_p(f)_T df. \tag{13}$$

If the process is ergodic, one can perform an ensemble average to obtain

$$<\Pi\,(T)>_e \;=<\Pi^j(T)>_e = \int_o^{\infty} < W^j_p(f)_T>_e df = \int_o^{\infty} W_p(f)_T df. \tag{14}$$

For an ergodic processes, the temporal average and expectation are equal,

$$<\Pi(T)>_{NT} \;=<\Pi(T)>_e. \tag{15}$$

The expected value of the power spectral density, as a direct consequence of the Wiener–Khintchine theorem (Middleton 1987), is related to the covariance function, K, and using Eq. (12),

$$\begin{aligned} W_p(f)_T &= 2\,F\{K\} = 2\int_{-\infty}^{+\infty} K\,exp(-i\omega\, t)dt \\ &=< (2/T)|F(p(T))|^2>_e \end{aligned} \tag{16}$$

The Fourier transform of the autocorrelation or covariance function is equal to 2/T times the spectral densities. Thus, the autocorrelation and covariance are equal to the inverse transform of the power spectral density.

$$\begin{aligned} &W_p(f)_T =< (2/T)|F(p(T)|^2>_e =< F\{\Gamma(\tau)\}>_e \\ &\rightarrow <\Gamma(\tau)>_e = (1/2)\int_{-\infty}^{+\infty} W_p(f)_T\, exp\,(+i\omega\,\tau)df = K(\tau). \end{aligned} \tag{17}$$

The response of measurement system shown in Fig. 4.1 to stationary statistical noise can be summarized by the power spectral density in relationship to the Fourier transforms of the autocorrelation, covariance, or pressure.

$$\begin{aligned} W(f)_T &= <F\{\Gamma(\tau)\}>_e = <F\{K(\tau)\}>_e \\ &= 2\pi\ <P(\omega)P(\omega)^*>/T = <P(f)P(f)^*>/T. \end{aligned} \tag{18}$$

The power spectral density level should be referenced to watts per square meter hertz as the natural units or when the power per unit area is proportional to pressure squared, the reference quantity is micropascals squared per hertz. The power spectral density level is defined as follows:

$$\begin{aligned} \text{Power spectral density} &\equiv 10 \cdot \log_{10}[W(f)_T/W_{ref}(f)_T] \\ &= 10 \cdot \log_{10}[(<p^2(t)>_T/\Delta f)/((\mu\,Pa)^2/Hz)]\ ;\ dB\ re\,(\mu\,Pa)^2/Hz). \end{aligned} \tag{19}$$

Some Simple Statistical Concepts

The texts *Detection of Signals in Noise* by Whalen (1971) and *Measurement and Analysis of Random Data* by Bendat and Piersol (1966) are excellent in their treatment of the ocean acoustics measurement problem. Although it is realistic to state that deterministic, transient, and random data analysis is an all-inclusive subject, it is the purpose here to simplify the treatment to the direct application of some simple statistics applied to ambient noise measurements. One needs to distinguish between time averages and ensemble averages, because they are only equivalent in the case of wide-sense stationary processes.

The simplest noise case is a stationary pressure time series that has a white spectral density, uniform with frequency, and is a zero mean Gaussian random process. As shown in Fig. 4.1, the time-varying pressure is converted to a time-varying voltage by the hydrophone and preamplifier, $p(f,t) \rightarrow v_i(f,t)$. The pressure is a zero mean Gaussian process and, likewise, so is the voltage, with the following probability distribution:

$$p_d(v_i) = \exp(-v_i^2/2\sigma^2)/\sigma\sqrt{2\pi}. \tag{20}$$

The voltage, $v_i(f,t)$, a Gaussian random variable, is filtered by a linear filter with a center frequency, $2\pi f_o = \omega_o$, and bandwidth $\Delta f[Hz]$. The linear filter output, $v_o(f_o, \Delta f, t)$, is a narrow-band Gaussian variable and when viewed on an oscilloscope is a slowly undulating cosinusoidal function of time:

$$\begin{aligned} v_o(f_o, \Delta f, t) &= v_{oi}(f_o, \Delta f, t)\cos(\omega_o t + \phi(t)) \\ &= (v_{oi}\cos(\phi(t))\cos(\omega_o t)\ +\ (-v_{oi}\sin(\phi(t))\sin(\omega_o t) \\ &= \alpha(t)\cos(\omega_o t)\ +\ \beta(t)\sin(\omega_o t) \end{aligned} \tag{21}$$

Since the sine and cosine are orthogonal or in quadrature, and since α and β are Gaussian variables, one usually treats these quantities as independent even though over short-time samples there is a phase for which they will be correlated. Consequently, the quantities are strictly not statistically independent random variables. However, the correlation of large numbers of sine and cosine samples decreases in inverse proportion to the number of samples and thus can be treated as statistically independent.

The voltage when squared and integrated over time T is

$$
\begin{aligned}
< v^2 >_T &= (1/T) \int_o^T v_o^2 dt = (1/T) \int_o^T (\alpha(t)^2 \cos(\omega_o t)^2 + \beta(t)^2 \sin(\omega_o t)^2) dt \\
&= (1/T)\,\alpha(t)^2 \int_o^t \cos(\omega_o t)^2 dt + (1/T)\,\beta(t)^2 \int_o^t \sin(\omega_o t)^2 dt \\
&= (1/2) \cdot (\alpha(t)^2 + \beta(t)^2).
\end{aligned}
\tag{22}
$$

In the above equation, cross-product terms integrated to zero. Here it has been assumed that the temporal variation of the Gaussian variables α and β is slow compared with the period of the filter's center frequencies and the integration variable T. With the introduction of a new variable $z^2 = 2 < v^2 >_T = \alpha^2 + \beta^2$, the probability density can be determined by standard techniques. Given that α and β are Gaussian,

$$
\begin{aligned}
p_d(\alpha) &= (1/\sigma\sqrt{2\pi})\exp(-\alpha^2/2\sigma^2) \ \ and \ \ p_d(\beta) = (1/\sigma\sqrt{2\pi})\exp(-\beta^2/2\sigma^2) \\
p_d(\alpha, \beta) &= p_d(\alpha) \cdot p_d(\beta) = (1/2\pi\sigma^2)\exp(-(\alpha^2 + \beta^2)/2\sigma^2)
\end{aligned}
\tag{23}
$$

To determine the probability distribution for the amplitude z and its square z^2, one uses the cumulative distribution function, P_c. If α and β are statistically independent random variables, the joint cumulative distribution function is

$$P_c(z \leq Z) = \iint p_d(\alpha, \beta) d\alpha\, d\beta. \tag{24}$$

With a change of variables, $\alpha = z\,cos(\theta)$, $\beta = z\,sin(\theta)$, with $z = (\alpha^2 + \beta^2)^{1/2}$ and $\theta = tan^{-1}(\beta/\alpha)$, the Jacobian of the transformation is $|J| = z$. The result is the joint cumulative function

$$P_c(z \leq Z) = \iint p_d(\alpha = z\,cos(\theta)) p_d(\beta = z\,sin(\theta))\, |J| dz d\theta. \tag{25}$$

Substitution of the probability density functions and integration over θ yields

$$
\begin{aligned}
P_c(z \leq Z) &= \int_o^Z \cdot \int_o^{2\pi} (1/2\pi\sigma^2) exp(-z^2/2\sigma^2) z dz d\theta \\
&= \int_o^Z \cdot (1/\sigma^2) exp(-z^2/2\sigma^2) z dz.
\end{aligned}
\tag{26}
$$

Differentiation of this expression produces the probability density of the magnitude z or the envelope:

$$p_d(Z) = (dP_c(z < Z)/dZ) = (Z/\sigma^2) exp(-Z^2/2\sigma^2), \ Z \geq 0. \tag{27}$$

The above probability distribution of the envelope or the root mean square voltage is recognized as the Rayleigh distribution. This distribution has an expected mean value, $E[z] = (\pi/2)^{1/2}\sigma$, and an expected second moment, $E[z^2] = 2\sigma^2$.

The procedure can be repeated to determine the probability distribution for the squared variable, z^2. First, let $u = z^2$ and use the fact $P_c(u \leq Z^2 = U) = P_c(z \leq Z)$.

$$\begin{aligned} P_c(z \leq Z) &= \int_o^Z [(z/\sigma^2)exp(-z^2/2\sigma^2)]dz \\ &= \int_o^{Z^2} [(1/2\sigma^2)exp(-z^2/2\sigma^2)]dz^2. \end{aligned} \tag{28}$$

In terms of the variable u,

$$\begin{aligned} P_c(u \leq U) &= \int_o^{Z^2} [(1/2\sigma^2)exp(-z^2/2\sigma^2)]dz^2 \\ &= \int_o^U [(1/2\sigma^2)exp(-u/2\sigma^2)]du. \end{aligned} \tag{29}$$

Differentiation of the cumulative distribution function yields

$$p(U = Z^2) = dP_c(u \leq U)/dU = (1/2\sigma^2)\ exp(-U/2\sigma^2). \tag{30}$$

This probability density can be written as follows:

$$p_d(z^2) = (1/2\sigma^2)\, exp(-z^2/2\sigma^2)\ , E[z^2] = 2\sigma^2\ ,\ E[z^4] = 4\sigma^4. \tag{31}$$

This is an exponential distribution or a two degrees of freedom χ^2 distribution. Thus, the description of the noise-processing scheme is complete. The postsummation of N independently sampled noise estimates is also χ^2 distributed with 2N degrees of freedom since the sum of χ^2 variables is also χ^2.

Phenomenological Observations

In *Ambient Noise in the Sea*, Urick (1984) commented on the prolific nature of the literature concerning theories of sound generation at the sea surface and measurements of the temporal and spatial spectral characteristics. The idealized spectral characteristic suggested by Urick was in agreement with the schematic of Wenz (1962) and for frequencies greater than 500 Hz is consistent with the observations of Knudsen et al. (1948) and Wenz's "rule of fives." The deepwater noise spectrum level can be described by

$$\begin{aligned} NL(f, U) &= 25 - 10 \cdot log[f^{5/3}] + 10 \cdot log[(U/5)^{5/3} \\ or& \\ NL(f, U) &= 25 - (5/3) \cdot 10 \cdot [log[f] - log[U/5]] \end{aligned} \tag{32}$$

where f is the frequency (kHz), U is the wind speed (knots), and NL is the noise level, which at 1 kHz and 5-knot wind speed is 25 dB re (20 $\mu N/m^2)^2$/Hz. He recommended the addition of another 5 dB for shallow water levels. Wenz observed in the 10–500-Hz band that the measured noise levels were often variable, dominated by shipping noise, with only weak wind-speed dependence. The shape of the spectrum also was found to vary from a positive slope to a steep negative slope. Below 10 Hz, he indicated a steep spectral slope of up to 10 dB/octave as the frequency decreased. These trends are shown in Fig. 4.2. Also shown in this figure are limits based on lake noise measurements, heavy rain, and the thermal noise limit due to molecular agitation. It is convenient to discuss this spectrum in nominal frequency

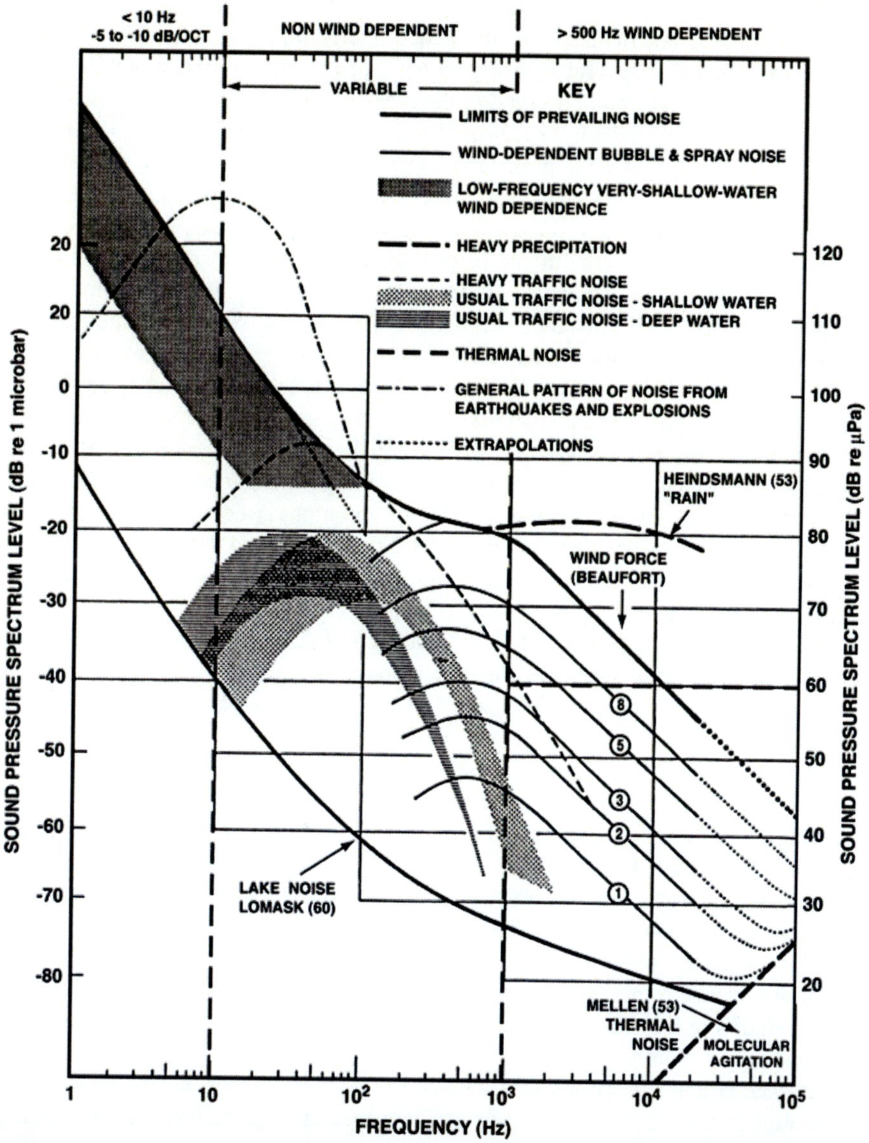

Fig. 4.2 The Wenz curves for ambient noise spectrum levels [adapted from Wenz (1962)]

regions. For example, the frequency region below 20 Hz is referred to as infrasonic; the band between 20 and 20 kHz is the sonic region; and that greater than 20 kHz is the ultrasonic region.

Infrasonic Noise

Infrasonic noise ($f \leq 20\,\text{Hz}$) was thought to be due to wave–wave interactions, capillary wave interactions, wave–turbulence interactions, and atmospheric turbulence. The results of several investigations since the time of Wenz are shown in Fig. 4.3. The general trend observed in the measured results is a decrease with frequency, 0.2–5 Hz, of the noise spectrum level of 5–12 dB/octave, dependent on the wind speed. These infrasonic experiments were difficult to conduct and temperature fluctuations and local flow noise can affect the measurement sensors. In addition, knowledge of a wide area of the bottom and the sea surface spectrum is required to define the experimental conditions. In general, the results shown in Fig. 4.3 agree qualitatively with the formulation of Miche (1944) and the further treatments by Longuet-Higgins (1950) insofar as the spectral peak is approximately double that of the surface wave spectrum and correlated with the surface displacement spectrum.

Other candidates for the production of the spectrum characteristics were the turbulent boundary pressure fluctuations, wave–turbulence interactions, and capillary wave interactions. Yen and Perrone (1974) and Goncharov (1970) placed estimates on these mechanisms and showed that below 5 Hz the estimated magnitude and spectral character are inconsistent with the observations, whereas above 5 Hz they provide a lower limit on the noise observed. The estimates of the wave–wave interaction by Hughes (1976) and Brekhovskikh (1966) appear to be consistent with the experimental results.

The measurements shown in Fig. 4.3, were complied by Kibblewhite and Evans (1985) and show that wave–wave interaction can the characteristic of the ambient noise spectrum at frequencies less than 5 Hz and the presence of the microseismic spectral peak in the 0.2 Hz. However, a definitive experiment that employs simultaneous measurement with seismic sensors, very low-frequency pressure sensors and wave spectrum instrumentation has yet to be performed. The work of Kibblewhite in extending the Wenz curve to the infrasonic region, although meritorious and timely, should be reexamined.

Cato (1991) extended the wave interaction theory and compared calculations with the results obtained from careful experiments in the lake of the Woronora Dam. Measurements in the 35-m-deep lake showed the expected agreement between the wave height spectrum and the noise spectrum. As a basis for comparison, Cato used a formulation of the wave–wave interaction using the Heaviside function to specify the sea surface boundary condition. This theoretical method is equivalent to the use of retarded Green's functions, as discussed in Chapter 3. The measured noise levels were higher than the open-ocean levels; however, the agreement with theoretical expectations for the specific conditions in his experiment was found to be significant. By extrapolation of these results, Cato concluded that there was no longer

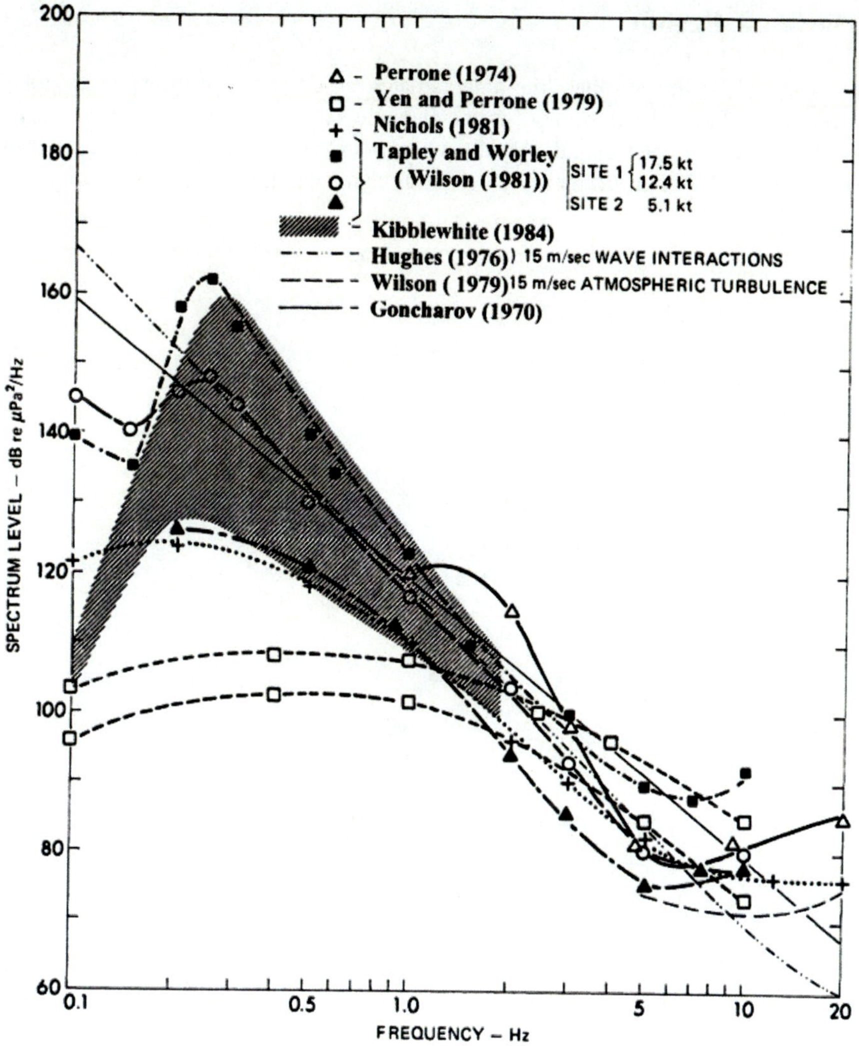

Fig. 4.3 Summary of infrasonic measurements and theoretical estimates [adapted from Kibblewhite and Ewans (1985)]

any doubt that wave–wave interactions produced the oceanic noise levels at frequencies between 0.1 and 5 Hz with an experimental spectral peak twice that of the surface wave height spectrum. The spectrum was observed to decrease steeply, probably a consequence of his experimental location. Nevertheless, the extrapolation to the open-ocean noise field was comparable to the estimates of Hughes, shown in Fig. 4.3.

The Sonic Region

Kerman (1984) showed that "the amalgamated observations of the ambient noise reveal a similarity structure, both in the acoustic spectrum and wind dependency." For frequencies greater than the local maximum in the 300–500-Hz region, Kerman found that the normalized measured spectral characteristic was proportional to f^{-2}

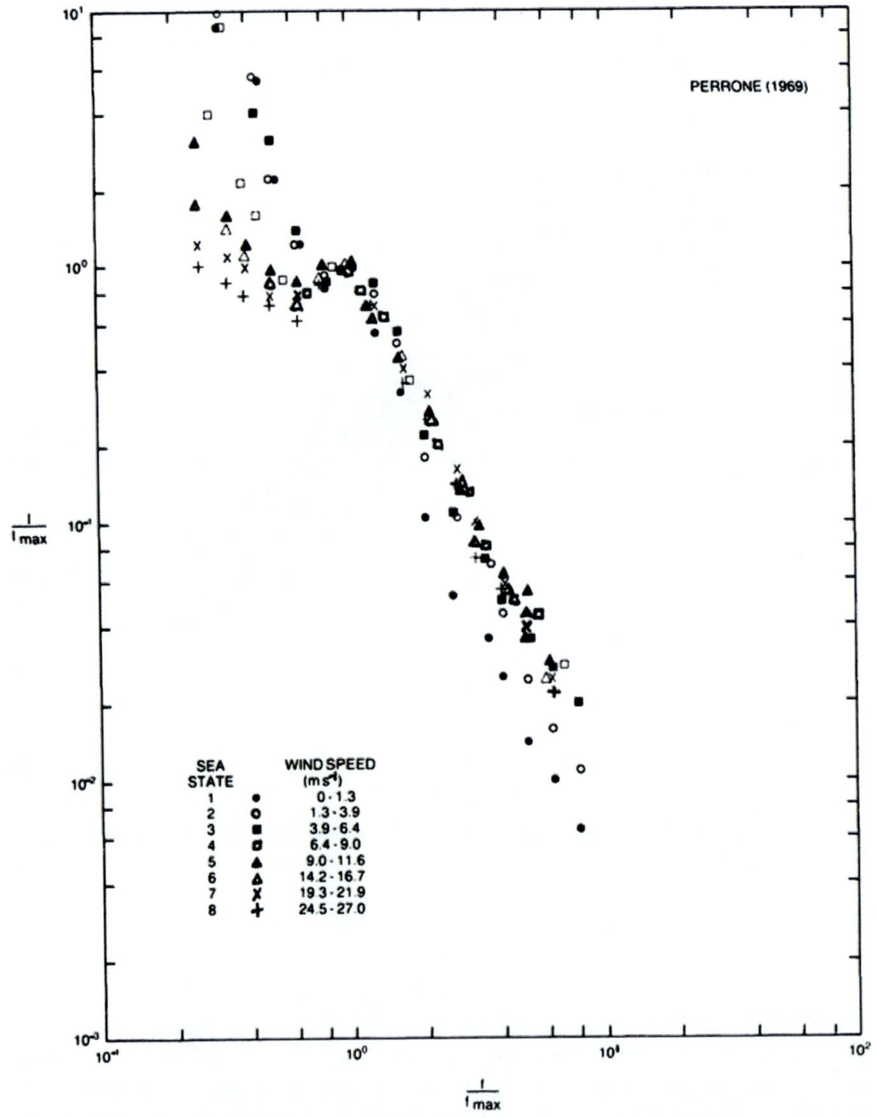

Fig. 4.4 Relative noise intensity versus relative frequency for the results of Perrone (1969) Kerman (1984) for a variety of wind speeds

(6 dB/octave), (Fig. 4.4). Furthermore, he showed that the noise intensity was proportional to the cube of the friction velocity (u_*^3) prior to a critical friction velocity (u_{*c}) determined by the minimum phase velocity of the gravity–capillary waves. Wave breaking was associated with this critical condition, and for $u^* > u_{*c}$, the noise intensity was found to increase with $u_*^{1.5}$ (Fig. 4.5) (see Chapter 2). The results shown in Figs. 4.4 and 4.5 represent a significant extension of Wenz curves for ambient noise levels.

Kerman concluded that the two distinct regions of ambient noise wind speed dependency existed, and suggested the presence of two sound source generation mechanisms or one mechanism that changed sensitivity. Since breaking waves were known to produce bubbles, spray, and splash, combinations of these mechanisms can explain the production of sound at frequencies greater than 500 Hz. Before wave breaking, bubbles can be produced by hydrodynamic surface instabilities and, consequently, one may have one mechanism that changes intensity since wave breaking produces more bubbles and consequently more sound. Kerman observed variability in the region below 500 Hz, consistent with Wenz.

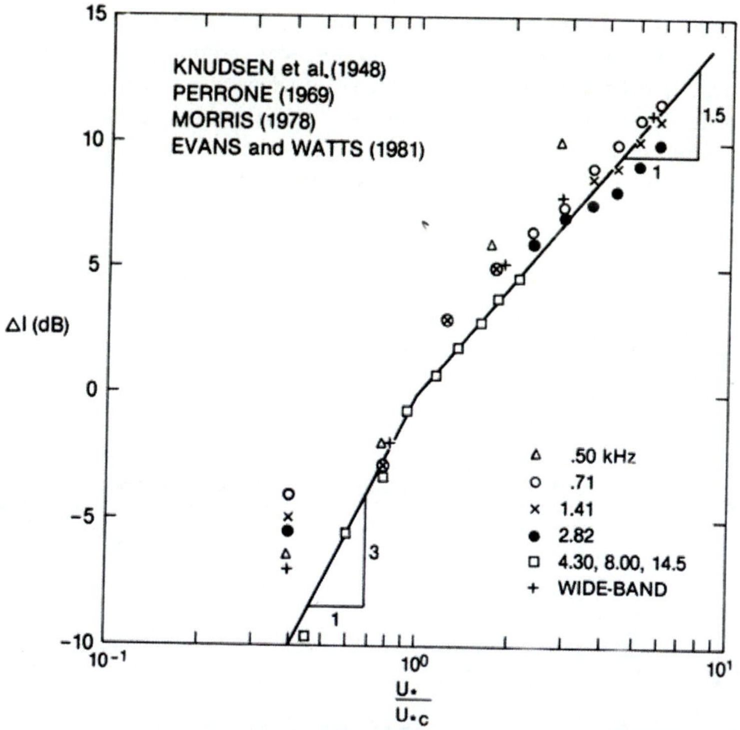

Fig. 4.5 The change in intensity level as a function of normalized friction velocity. Prior to the critical friction velocity there is one wind dependence, and for friction velocities greater than critical, there is another. The critical friction velocity is determinative of wave breaking Kerman (1984)

Before wave breaking, Kerman showed that the noise intensity was proportional to the cube of the friction velocity (u_*^3) prior to the critical friction velocity (u_{*c}), and for $u^* > u_{*c}$, the noise intensity was found to increase with $u_*^{1.5}$.The result is shown in Fig. 4.5 and the two wind speed regions are important to the estimation of ambient noise source level curves.

At frequencies less than 500 Hz, it was difficult to acquire ambient noise measurements attributable solely to local sources. The reasons for this difficulty were the dominance of noise radiated by shipping, the problems of measurement system self-noise, the size required for directional measurement, and the relatively low long range propagation loss. Figure 4.6 shows a comparison of measured noise spectrum levels ranging from 1 to 400 Hz, the very low frequency to high-frequency region. The observations presented in this summary are consistent with the observations of Wenz insofar as the region between 1 and 5 Hz shows a steep spectral slope (−5 to −10 dB/octave) with a wind speed dependence and a region above 100 Hz with a weak wind speed dependence, except for the measurements by Frisch in the Mediterranean Sea. The region between 5 and 100 Hz is basically wind-independent and the variation in levels is apparently determined by basin size and shipping density.

However, very low frequency to mid-frequency (1–500-Hz) ambient noise measurements that were not dominated by noise radiated from shipping at these frequencies showed the presence of locally wind-generated noise. Wittenborn (1976) found wind speed dependence with linear velocity dependence prior to u∼6 m/s and a nonlinear dependence at greater wind speeds. The results are shown in Fig. 4.7 for the frequency range of 1–500 Hz. The experiment consisted of a vertical string of omnidirectional hydrophones; one was positioned below the sound channel's

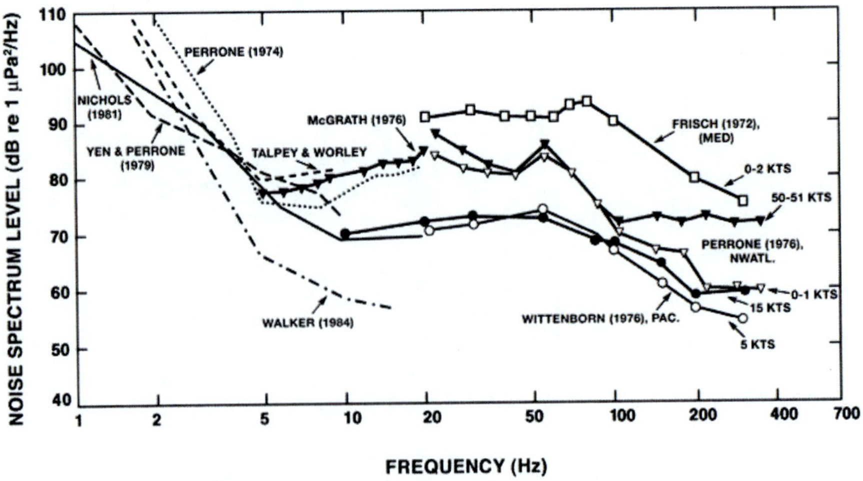

Fig. 4.6 Comparison of noise spectrum levels in the frequency range from 1 to 400 Hz in several ocean basins

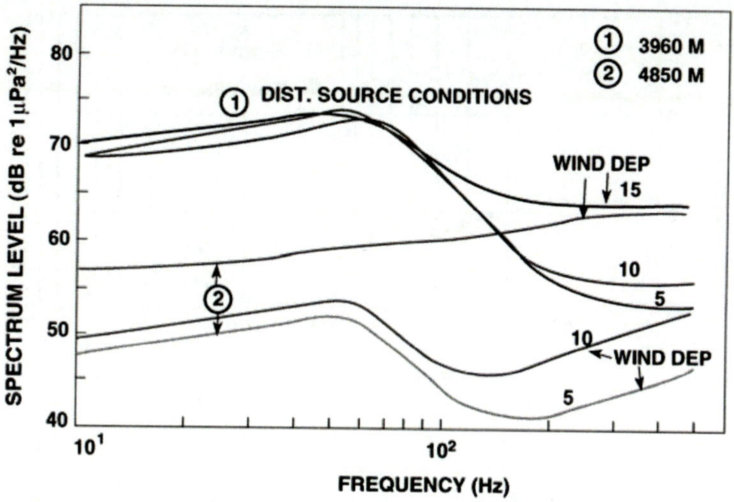

Fig. 4.7 The noise spectrum level from 1 to 500 Hz for a near-bottom hydrophone [adapted from Wittenborn (1976)]

critical depth to minimize the hydrophone's reception of distant generated noise, whereas the other was positioned in the sound channel. [Results were republished by Shooter, De Mary, and Wittenborn (1990).] The hydrophone data were recorded remotely on a time-indexed magnetic tape and wind speed estimates were based on interpolating weather reports of ships transiting the area. Stationary periods were selected by examining the continuous recording of the 300–500-Hz band of the deep hydrophone. The results clearly show wind-dependent noise characteristics between 1 and 500 Hz. The important feature, shown in Fig. 4.7, is that before wave breaking (wind speed about 10 knots), the ambient noise levels on the deepest hydrophone have a characteristic consistent with noise leakage from the sound channel.

After wave breaking occurs, the noise level measured by hydrophones below the critical depth for a wind speed of 15 knots is increased by 15 dB over the 1–500-Hz band. The hydrophone signal at the sound channel depth was still dominated by the noise from distant ships. This observation shows low-frequency sound is produced in breaking wave events. The low-frequency sound from breaking waves breaking is most likely due to microbubble plumes and clouds associated with breaking waves.

Observations in the sparsely shipped Southern Hemisphere represent measurements not dominated by shipping. Kibblewhite (1976) compared the levels between the North Pacific and the South Pacific as shown in Fig. 4.8. The relative difference in levels is striking between 15 and 200 Hz. Cato (1976) examined 40 sites near Australia and found wind speed dependence between 20 and 5,000 Hz. His measurements also included the effects of moderate shipping and considerable biological activity.

Measurements performed with vertical arrays in the sparsely shipped Southern Hemisphere's Fiji Basin [Bannister (1981), Burgess and Kewley (1983), Browning

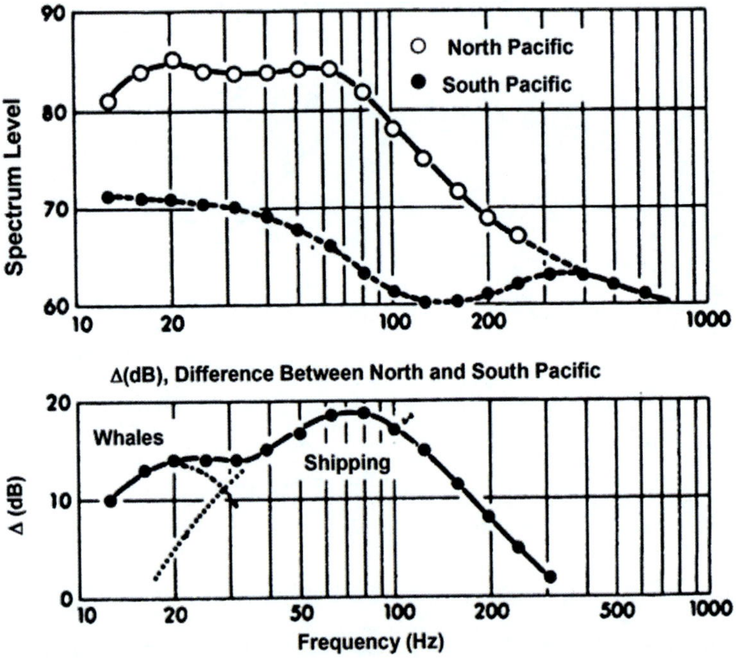

Fig. 4.8 Comparison of measured noise levels in the North Pacific and the South Pacific. [Kibblewhite (1976), note the ordinate label should be "Spectrum level $dB\ re\ (\mu Pa)^2/Hz$."]

(1986)] were also found to have low-frequency wind noise with two distinct wind speed dependencies. When compared with other shallow water and deepwater noise measurements, these results led to the conclusion that at lower frequencies sound was generated near the sea surface with one wind-speed-dependent mechanism prior to wave breaking and another after wave breaking. Since field observations are dependent on the mechanism of sound production at the sea surface and sound transmission is determined by environmental factors such as water depth, bottom properties, and the sound velocity structure, the measured noise levels can be variable. The conclusion, however, was that in the low- to high-frequency region, the noise characteristics were determined not only by propagation but also by whether shipping or breaking wave sources of sound are dominant. The question posed was: How does a breaking wave produce low-frequency sound?

Two interesting experiments have been performed featuring visual as well as acoustic identification of wave breaking events. Hollett (1988) performed an experiment in the Mediterranean Sea with a vertical array of hydrophones (three nested apertures (32 phones each) with center frequencies at 375, 750, and 1,500 Hz) and simultaneous video recording of the sea surface area intersected by the endfire beam of the array. Figure 4.9 shows the spectral events resulting from a large spilling breaker. The spectra shown have not been corrected for the prewhitening of

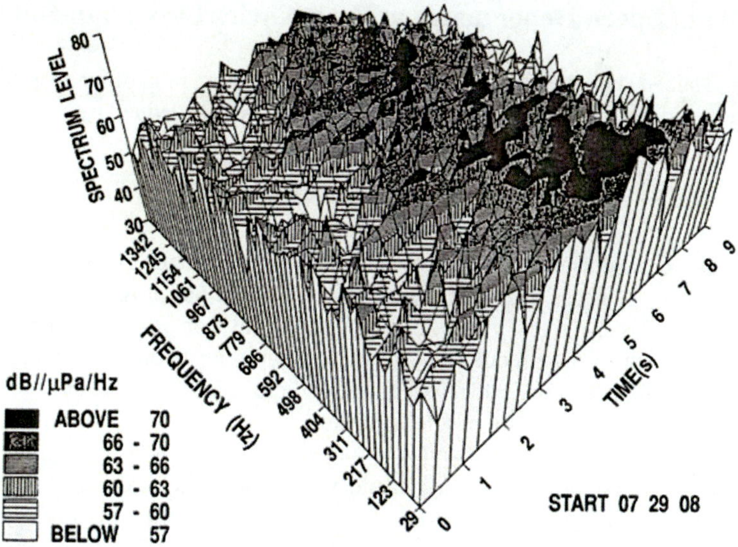

Fig. 4.9 Spectra of a wave breaking as measured by Hollett (1992) as a function of time. The lower-frequency spectral estimates have not been corrected for pre-emphasis. This collection of acoustic spectra shows the wave-breaking event to be a random collection of spectral peaks

the data; i.e., the lower-frequency spectral content is more pronounced than shown. The breaking event occurs between 3 and 7 s and is seen to be a random collection of spectral peaks most pronounced in the low-frequency (below 300 Hz) range.

Thus, a plausible explanation is that at higher frequencies single bubble noise is important. At lower frequencies, the collective oscillations of regions with sufficient compressibility are responsible for the sound. These radiations are described by the oscillations of these regions as if they were single radiators and have frequencies described by a modified Minnaert equation (see Chapter 3 or Appendix D). Farmer (1989) performed an experiment in 200 m of water with a hydrophone 14 m below the surface and was able to identify the occurrence of wave breaking by examining the video obtained from a subsurface camera. The simultaneously recorded acoustic data were examined and found to show that the breaking waves radiate sound to frequencies as low as 50 Hz.

The Hollett and Farmer results show that breaking waves are a source of low-frequency sound and that at frequencies below 500 Hz the ambient noise spectrum in the absence of shipping is wind-speed-dependent. These results are important with respect to measurements performed with vertical and horizontal arrays, where the directional noise properties, in the absence of shipping, are due to the interaction of wind-generated noise and the basin boundaries.

The Wind Speed Dependency at Frequencies Less Than 500 Hz

The wind speed characterization was based on the determination of two parameters, the spectral slope m(f) and the wind speed dependency n(f) defined by

$$\Delta NL(f, f_o U, U_o) = 20n(f)\mathrm{LOG}(U/U_o) + 10m(f) \cdot \mathrm{LOG}(f/f_o), \qquad (33)$$

where ΔNL is the difference in measured noise level at frequencies f and f_o for wind speeds U and U_o. These parameters are estimated by the use of regression analysis and, in some instances, by visual least-squares estimation. As stated previously, the measurement of low-frequency wind-dependent noise is difficult owing to the presence of shipping noise, self-noise, and sound propagation conditions.

Shown in Fig. 4.10 is the problem of determining n(f) in the case of a two-component noise field, shipping- and wind-dependent surface noise. When the curve is horizontal, n = 0, no wind speed dependency, shipping noise dominates. At low frequency, n(13 Hz) = 2.10, for higher wind speeds n(283 Hz) = 1.36, and then n decreases to n(2.2 kHz) = 1.20.

The frequency-dependent characteristic of n(f) was observed in other experiments (Fig. 4.11). n(f $\leq$ 40 Hz) $\approx$ 2.6, but n(f $\geq$ 300 Hz) $\approx$ 1.2.

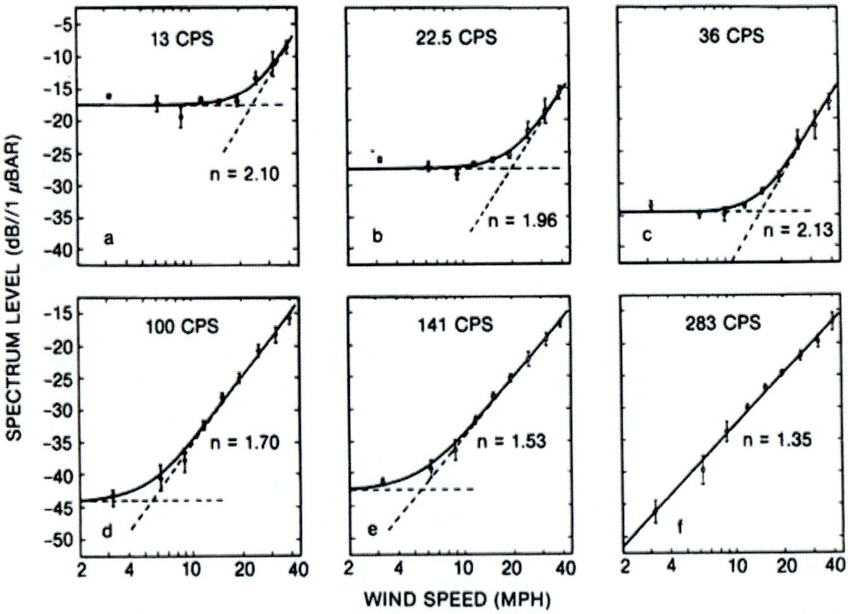

Fig. 4.10 Frequency-dependent wind speed characteristic for several frequencies. The *horizontal lines*, no wind speed dependence, at lower frequencies and low wind speed show the limiting effect of noise from distant shipping. [(Piggott 1964), note the ordinate label should be "Spectrum level *dB re* $(\mu Pa)^2/Hz$" and is obtained by adding 100 dB.]

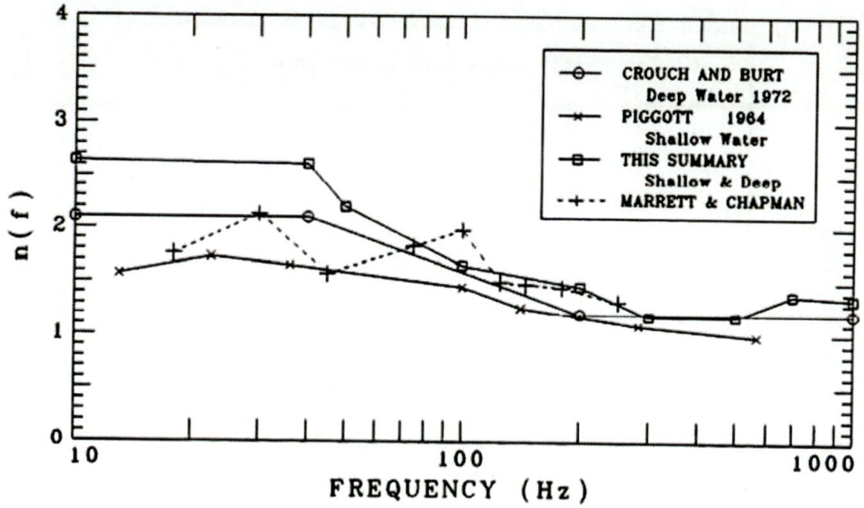

Fig. 4.11 Wind speed dependency factor n(f) versus frequency for wind speeds above 13 knots. Shown are the results of Crouch and Burt (1972), Piggott (1964), and Marrett and Chapman (1990), and an average of all results

In the previously discussed Wittenborn (1976) experiment, a vertical string of omnidirectional hydrophones found a wind speed dependence with a linear velocity dependence prior to u~6 m/s and a nonlinear dependence at greater wind speeds. The 15-knot curve from this experiment, Shooter et al. (1990), in Fig. 4.12 has spectral slopes of m = 0, m = 1, and m = 1/3. The corresponding wind speed factor (n), determined by the difference in noise levels at 12 and 15 knots for a frequency of 100 Hz, was n = 1.65.

Several investigators have used vertical arrays to measure the locally generated noise in deep water. Most notable are the works of Burgess and Kewley (1983) in the South Fiji Basin and Australian waters, and Kennedy (1990) in the Tongue of the Ocean near the Bahamas. In addition to the use of steerable arrays, these experiments were conducted in areas to minimize shipping noise and had concurrent measurement of wind speed. The Burgess and Kewley (1983) experiment used a 180-m steerable array at a depth of 300 m in deep water to avoid mixed-layer ducting effects. Upward- and downward-looking beams were used to estimate the source level (dB *re* 1 $\mu Pa^2/(sr \cdot \text{Hz})$) at the sea surface. Since their original publication, these authors have reexamined their linear regression analysis of the wind speed dependency and have concluded that two wind-speed-dependent regions exist, i.e., one prior to and one after the onset of wave breaking with n~1.5 (Kewley et al. 1990).

The frequency-dependent source level estimate is shown in Fig. 4.12. In examining Fig. 4.12, one must differentiate between noise level and source level estimates that have been plotted on the same scale to show that there is no strong spectral dependence at wind speeds greater than 6 m/s (12 knots). In the case of the Kewley (1990) results, we observe at most m = 1/3 between 100 and 200 Hz.

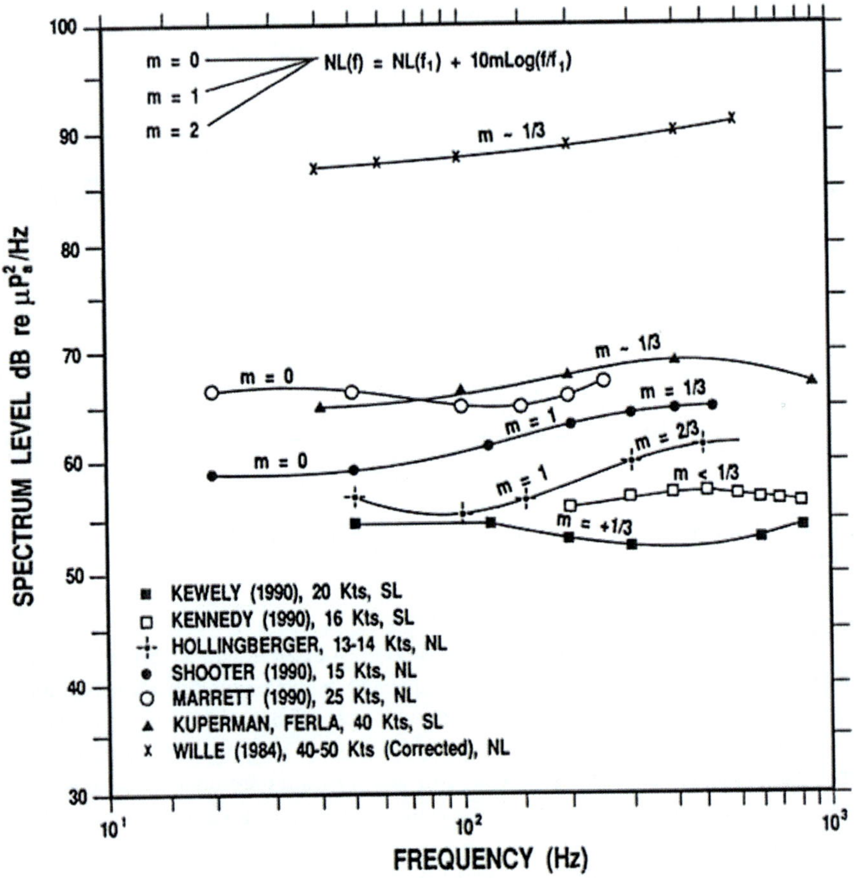

Fig. 4.12 Source spectrum level (SL) and noise spectrum level (NL) versus frequency for wind speeds above 13 knots for several diverse oceanographic areas

The Kennedy (1990) experiment was performed with a measurement array of seven-octave nested four-wavelength linear apertures covering the 40–4,000-Hz band. Meteorological measurements were performed on the nearby Andros Island. The measurements were made at a considerable height above the land surface of the island and were used to estimate the critical speed at the ocean surface at the measurement location. In his experiment, Kennedy was able to model the propagation in the Tongue of the Ocean. The vertical distribution of ambient noise N(θ) can be written as (Talham 1964)

$$N(\theta) = dI/d\Omega = DPWg(\theta')\exp(-2\alpha r)/\cos\theta'(1 - \beta\gamma\exp(-4\alpha r)), \qquad (34)$$

where $W = [\sin\theta'/\sin\theta] = [C_s/C_r]^2$, D is the source density at the surface (number/m^2), P is the power (W/sr), $dI/d\Omega$ is the intensity per unit solid angle

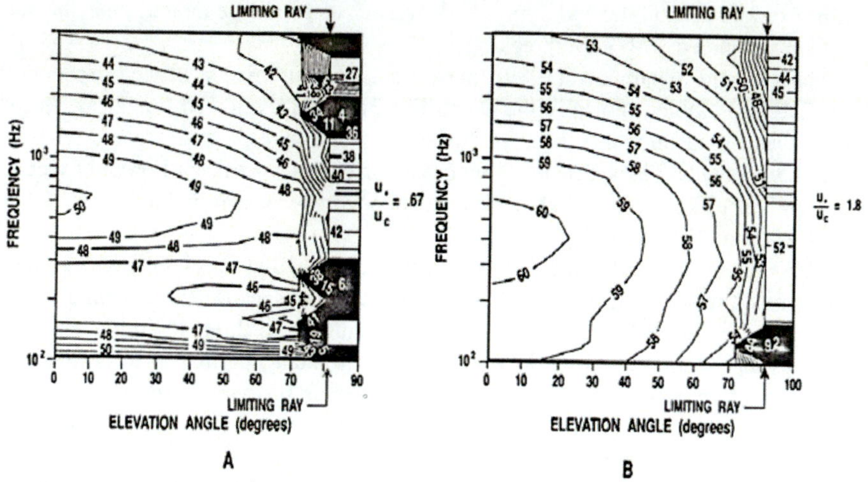

Fig. 4.13 Estimated noise spectrum level versus elevation angle as estimated by Kennedy (1990) for **a** before wave breaking and **b** after wave breaking

(W/m^2/sr), α is the frequency-dependent absorption coefficient, γ is the surface reflection coefficient, β is the reflection coefficient of the bottom, θ' is the source angle, θ is the angle at the receiver, and C_s and C_r are the speed of sound at the source and the receiver, respectively.

Thus, given knowledge of the environmental factors, one could estimate $g(\theta')$ given a measurement of $N(\theta)$. Kennedy measured $N(f, \theta)$ and determined, with a similar but more complicated method of curve fitting, the best fit to$D \cdot P \cdot g(\theta')$. Kennedy's results (Fig. 4.12) for a wind speed of 8 m/s (16 knots) show a practically white spectral curve. The novel aspect of Kennedy's experiment was his ability to estimate the source directional characteristic $g(\theta')$ as a function of wind speed and frequency. Shown in Fig. 4.13 are two examples of $N(f, \theta)$ from this estimation based on his measured data. Figure 4.13a shows the estimated pattern at low wind speeds prior to whitecaps being present. A broad local maximum is observed at 600 Hz that decreases with an increase of elevation angle (0° downward), vertical directionality. At lower frequencies, the level is rather constant with elevation angle; no directionality was observed. Kennedy found with whitecaps present (Fig. 4.13b) a broad downward-directed maximum at 400 Hz consistent with the vertical directionality from a surface distribution of dipole or doublet sources over the entire frequency range.

The structure shown in Fig. 4.13a was modeled by dipoles near the broad maximum at 600 Hz, but at lower frequencies (below 300 Hz) the structure was consistent with distributed sources or noise from a distance. The structure presented in Fig. 4.13b was found to be consistent with dipole or doublet distributed sources, monopoles beneath the pressure-release surface. Kennedy's results show a slowly varying spectrum level with frequency, but also vertical directionality consistent with a dipole or doublet radiation characteristic of sound generated near the sea surface associated with wave breaking.

Marrett and Chapman (1989) performed measurements of low-frequency (15–250-Hz) ambient noise in the South Fiji Basin with a towed line array. Local wind speed measurements were performed. Shown in Fig. 4.12 are the 25-knot noise level results. The spectrum again has a fairly white character. For wind speeds greater than 15 knots, the wind speed dependency factor was found to be $n \simeq 1.32$ at 250 Hz and $n \simeq 2.13$ at 30 Hz.

These deepwater ambient noise measurements show that for wind speeds greater than 6 m/s, the sea surface sound has a slowly varying spectral shape with a broad maximum ranging from 300 to 500 Hz. The results from three additional shallow water experiments are shown in Fig. 4.12. Ferla and Kuperman (1984) reported results on the wind-generated source spectrum levels for wind speeds between 10 and 40 knots. They used propagation loss measurements and environmental information in a wave-theoretic noise model to derive these source levels from measured noise levels. As can be seen, their results are consistent with the deepwater curves. Also shown are results from Wille and Geyer (1984) in the North Sea at 40–50 knots and Hollinberger and Bruder (1990) in an Alaskan fjord at 13–14 knots. To obtain the curve attributed to Wille, an estimated but relative correction factor, based on measured propagation loss, was applied to the measured noise levels. All these results have similar spectral slope with frequency but different spectrum levels.

This characteristic of a broad maximum between 300 and 500 Hz has been recognized in wind-generated ambient noise for quite some time (Wenz 1962), (Piggott 1964), but was not identified with such a diverse set of experiments. The broad maximum between 300 and 500 Hz characteristic of the curves shown in Fig. 4.12 has a decreasing spectrum level as frequency decreases. This decrease corresponds to a spectral slope factor of $m \sim 1/3$ or 1 dB/octave. This characteristic has often been compared to the relative spectrum of noise due to a spray of water droplets developed by Franz (1959). However, the Franz spectrum has a steeper (1.3–1.4 dB/active octave) slope than the spectra here; nevertheless, the observation of their similarity is worthy of detailed consideration.

Mid- to High-Frequency Wind Speed Dependence

In the previous discussion of low-frequency noise measurements, the wind speed dependency characterized by the factor n(f) was observed to range from 1.3 to 2.5. Wind speed has traditionally been used to provide an index for the level of noise to be expected at higher frequencies. But even at high frequencies, the wind speed dependency factor may be variable and complex. To illustrate the complexity of the wind speed dependency factor n, we have produced the schematic shown in Fig. 4.14. The data shown were obtained in the North Sea, a region with appreciable shipping-generated noise. The specific data set illustrates what was found in the literature concerning the wind speed dependency. The spectral curve at 20 kHz below

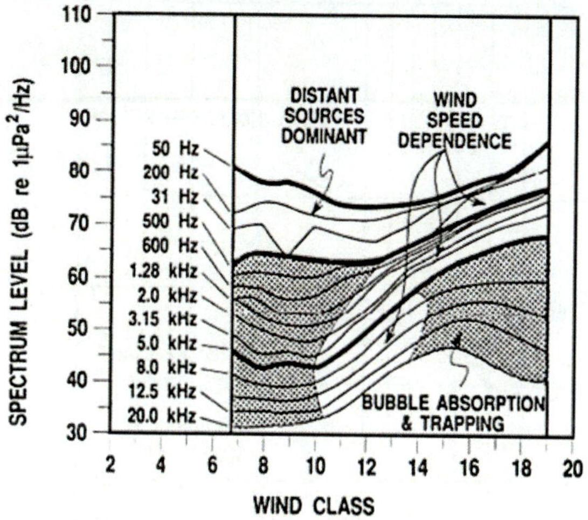

WIND CLASS	WIND SPEED
7	1.6 - 2m/s
8	2 - 2.5m/s
9	2.5 - 3.15
10	3.15 - 4
11	4 - 5
12	5 - 6.25
13	6.25 - 8
14	8 - 10
15	10 - 12.5
16	12.5 - 16
17	16 - 20
18	20 - 25
19	25 - 31.5
20	31.5 - 40

WD = 30m
WSH = 46m

Fig. 4.14 Spectrum level of ambient noise versus wind speed class in the North Sea. These spectra show the general wind speed dependency of ambient noise. [Adapted from Wille and Geyer (1984)]

a wind speed class of 10 shows no dependency ($n = 0$); between wind speed classes 11 and 13, there is a dependency of $n = 2$; and at higher wind speeds, there is a reversal of slope and a decrease in noise level.

This trend is seen to persist in some degree down to 5 kHz. The decrease in sound level at wind speeds greater than 10 m/s has been attributed to sound absorption by the near-surface microbubble layer and trapping in the near-surface sound duct produced by sound speed gradients resulting from the presence of microbubbles (Wille and Geyer 1984), Farmer and Vagle 1989).

At 1 kHz, we observe a region of no wind speed dependence below a wind speed class of 12 and a wind speed dependency of $n = 1.75$ at greater wind speeds. At lower frequencies, about 200 Hz, a wind speed dependency is only observed for higher wind classes (wind class above 16, wind speed above 12.5 m/s). In general, the observed wind speed dependence was a noise-limited region, a transition region, and a high-wind-speed region. These regions are shown in Fig. 4.15 for representative wind speeds and values of n. Classifying measurements in these regions may aid in the interpretation of low-frequency noise levels and may explain the variation found in the published values of $n(f)$.

This scheme has been used to examine the value for $n(f)$ for noise results obtained at wind speeds greater than 6.5 m/s for selected experimental results. The summary of this analysis is shown in Fig. 4.11 along with the values reported by Crouch and Burt (1972), Piggott (1964), and Marrett and Chapman (1990). The trend is clear for high winds speeds; n(f) has a constant value of about 2.5 until a frequency of 50 Hz and about 1.2 for frequencies greater than 300 Hz. The variation of n(f) with frequency between 50 and 200 Hz appears to be real. Kerman (1984) observed that all high-frequency results (above 500 Hz) yield a consistent set of characteristics.

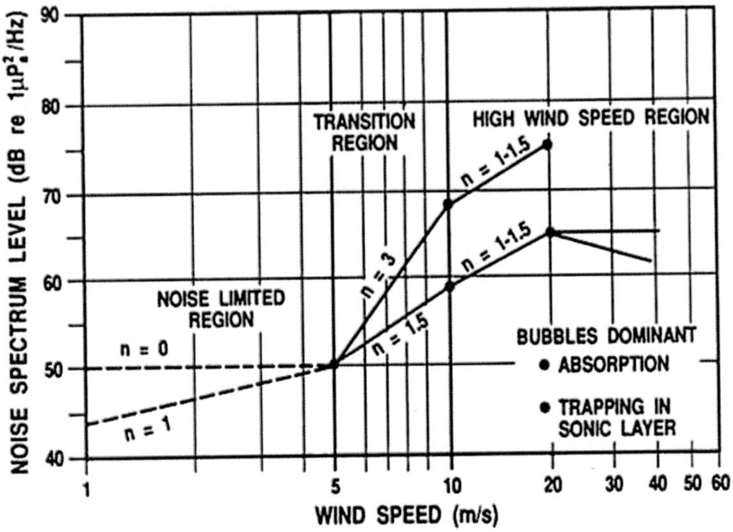

Fig. 4.15 Wind speed dependency ($n(f)$) versus wind speed for frequencies below 20 kHz. Three regions are identified: a noise-limited region, a transition region, and a high-wind-speed region

This indicates that sound production at these higher frequencies may be from the same mechanisms, such as bubble, splash, and spray. However, the variation of $n(f)$ at lower frequencies indicates a different mechanism before and after wave breaking.

Low-Frequency Noise Characteristics

One now has an answer to the question "What are the characteristics of low-frequency (below 500 Hz) ambient noise?" First, the cited evidence showed breaking waves produce sound with frequency as low as 30 Hz. The Hollett spectra appear to be a random collection of spectral peaks spread in frequency and time during the breaking event. These results are similar to those of Farmer (1989), although in some instances Farmer observed a broadband event representative of impact noise at the beginning of the wave breaking, followed by the subsequent random collection of spectral peaks.

Second, we showed that measurements of local wind-generated noise at wind speeds greater than 6 m/s have a broad maximum between 300 and 500 Hz. The position of this maximum shifts to lower frequencies as the wind speed increases. The decrease in spectrum level with a decrease in frequency is less than $m = 1/3$ (1 dB/octave) below this broad maximum.

Third, we showed that the wind speed dependency ($n(f)$) of local wind-generated noise is a complex function and may be characterized by three regions. In the high-wind-speed case, the low-frequency noise level (dB *re* 1 μPa^2/Hz) increased with wind speed dependency $n(f) = 1.5$ at 200 Hz, which ranged between 1.3 at 500 Hz and 2.5 at 30 Hz.

Finally, the results of Kennedy (1990) clearly show that low-frequency sound produced at the sea surface after the occurrence of wave breaking has a dipole radiation characteristic. Thus, the problem is reduced to finding a sound-generating mechanism that produces a broadband event followed by a random collection of spectral events, which scale according to $U^{2n(f)}$ power.

Directionality

The directionality of the oceanic noise depends on the source of the sound and the frequency-dependent propagation factors. Sources of sound near the surface of the sea have a dipole characteristic as discussed in Chapter 1 and Appendix G. Bubbles and bubble clouds are subsurface monopoles, and because of the surface image effect as well as the near-surface bubble-layer attenuation are always essentially "dipolelike." That is to say, at lower frequencies, oscillating bubble clouds caught in the orbital motion of the gravity wave field have a dipole characteristic. In the frequency region where single bubbles dominate the noise, one has a dipole characteristic at low wind speeds due to the image effect, whereas at higher wind speeds the combination of the image effect and the exponentially distributed bubble layer produces radiation patterns in the downward direction that can be considered "dipolelike." Furthermore, spray and splash of breaking waves act as dipole sources of sound.

Propagation of sound in the ocean determines the range at which noise sources are important. In both deep and shallow water, noise radiated at the surface in the high-frequency range is attenuated by the absorption in the seawater and boundary interactions, the rough sea surface and bottom. At lower frequencies, the attenuation decreases, the range of influence becomes larger, and reflection loss from the bathymetry is less. For example, propagation of sound near 50 Hz ensures that the reflection from slopes and seamounts is important because of the high bottom reflectivity. Owing to the angle of the slope, the downward-directed energy from the source of sound will reflect from the boundary with a reduction in angle of twice the slope angle. This reduction of the angle of propagation couples sound over bathymetric features into the sound channel. The consequence is a slope enhancement, the megaphone effect, and is profound for both surface-ship- and wind-driven noise at lower frequencies.

Low-Frequency Directional Noise Characteristics

Omnidirectional measurements of mean square pressure and its logarithmic counterpart, the ambient noise level, as stated previously, were usually made with a filter, a squaring device, an integrator, and an incoherent sum or averaged. Recent measurements were performed by digitizing the noise signal and taking the linear average of the FFT magnitude squared and then sequentially averaging the results. Both methods yield sample measures of the mean square pressure, the mean intensity, and variance that can be converted to an ambient noise level. The omnidirectional hydrophone spatially integrates the multiple sources of noise, the number of which, in any given time period, can be large and variable. The mean ambient noise levels can vary by as much as ±6 dB depending on the processing and averaging actually employed; however, the mean ambient noise levels are predictable from location to location with wind speed and shipping densities as controlling parameters.

Array processing can be accomplished either in the time domain or in the frequency domain. In the time domain, the hydrophone signals are amplified, summed after a suitable steering time delay has been applied, and then filtered, squared, and averaged as before. Shading may be applied to control side lobes. In the frequency domain, the signals are Fourier-transformed in the temporal and spatial domains to yield an equivalent beam power. Thus, one might expect levels analogous to the omnidirectional levels because the broad beams still contain many sources and multiple paths.

On the other hand, high-resolution arrays that resolve individual noise sources have dramatically different spatial and temporal characteristics. In the omnidirectional case, the mean square pressure is the squared sum of a large number of random components, whereas in the resolved case, the observable is the squared superposition of individual coherent components and a lower-level random component. The measurement of the directional characteristics of the noise field is a development since the work of Wenz.

Beam Noise Levels and Horizontal Directionality

The measurement of horizontal noise directionality (beam noise level versus angle) is at best difficult since measurements performed with horizontal line arrays have a left–right ambiguity. Because of this ambiguity, the measurement of the horizontal directionality requires either a deconvolution technique on the array beam-former output or the use of bathymetric shielding.

Wagstaff (1978, 1981, 2005) developed such a technique and used it to characterize the noise directionality in several ocean basins, such as the Northwest Atlantic, the Northeast Pacific, the Indian Ocean, and the Mediterranean Sea. The technique was approximate and required a directional array towed in a hexagonal pattern. The method involved iterative convolutions between an estimated field and the measured patterns until a least-squares criterion was achieved. The technique required approximately 20 h to perform and, thus, the estimated directionality was termed the "persistent component" of the low-frequency noise field. The hypothesis of Wagstaff was that the shipping densities, propagation of sound, and the bathymetry determine the directionality, and that this was a characteristic of each basin. In particular, he was mindful of the "megaphone effect" (6 dB) and concluded that surface ships over the continental slope were very important since shipping lanes usually proceed from the deep ocean to ports in a predictable pattern. This did not imply that fishing, continental shelf activity, the presence of oil wells, and exploration activity were ignored.

This approximate deconvolution method was evaluated in several sea tests during which the shipping density as a function of latitude and longitude was determined by over flights, the range-dependent sound propagation was measured, and the meteorological conditions and oceanic variables were known.

For frequency bands centered on 50 and 150 Hz, Fig. 4.16a shows results from a location near the Corner Seamounts in the Northwest Atlantic Basin, whereas Fig. 4.16b shows the results for the Gulf of Mexico for an array towed off the Yucatan Peninsula. For the Northwest Atlantic Basin, observe the 7.6–15-dB anisotropy, with the highest levels in the north–northeast, north and northwest; that is, in the direction of the major shipping lane (north, north-northwest), the Scotian Shelf and Halifax Harbor (north-northwest), and Grand Banks (north-northeast).

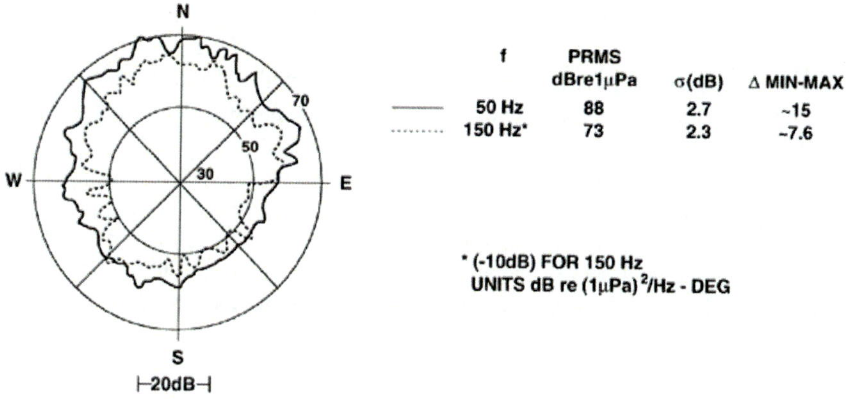

a.) The Northwest Atlantic

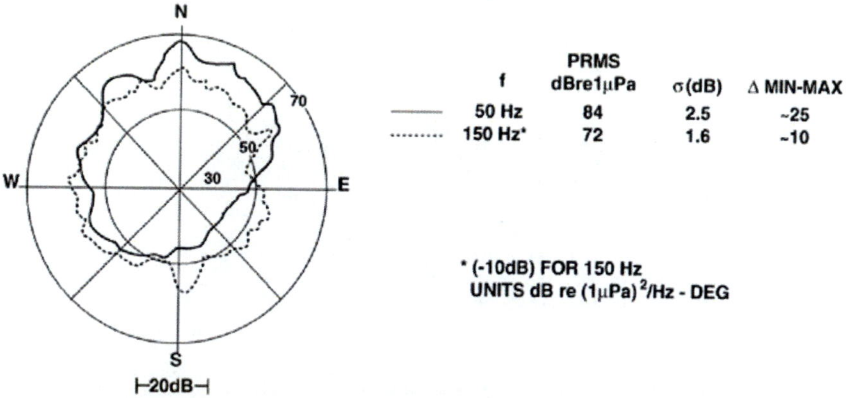

b.) The Gulf of Mexico

Fig. 4.16 Horizontal noise directionality: (**a**) the North Atlantic and (**b**) the Gulf of Mexico

The observed directionality was attributed to shipping, both mid-basin and traversing the continental rise shelf and Grand Banks. Also shown in this figure are the omnidirectional noise levels as well as the standard deviation derived from measured and estimated beam levels used in the estimation procedure.

The estimate of the horizontal directionality for the Gulf of Mexico showed high levels in the northern quadrant and low levels in the southern quadrant in the direction of Campeche Bank. Again, the difference between high and low levels was found to be 10 dB at 150 Hz and 25 db at 50 Hz, with pooled standard deviations of 1.6 and 2.5 dB, respectively. The northern quadrant contains the ports of Galveston and New Orleans, and oil-platform activity. Wagstaff showed that this long-time averaged pattern, the persistence, was correctly described by calculations provided that the shipping densities were known, transmission was calculable, and that ships over the continental shelves and seamounts were included. He showed that the megaphone effect coupled with shipping dominated these horizontal directional patterns.

Beam noise level time series were also obtained with an array using a plane wave beam-former and continuously recorded hydrophone signals in the Levantine Sea (4 h) and the Ionian Sea (10 h).

Figure 4.17 shows an example of the Levantine Sea beam noise levels versus the azimuth angle and time for a frequency of 320 Hz, a 110-wavelength aperture, a bandwidth of 0.18 Hz, and a sample time of 8 s, with the grayscale span of 40 dB normalized to the peak level. The shipping traffic, which is the marked by darker lines, is well tracked. This aperture resolves the shipping and provides a contrast

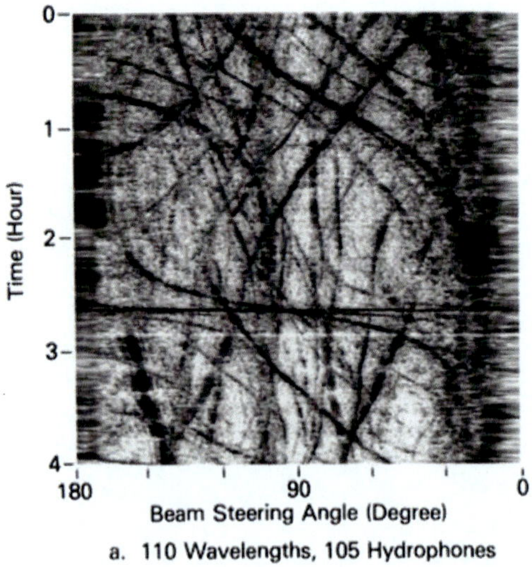

Fig. 4.17 Beam noise intensity levels versus time and steering angle for a frequency of 320 Hz, 0.18-Hz bandwidth, and 8-s interval

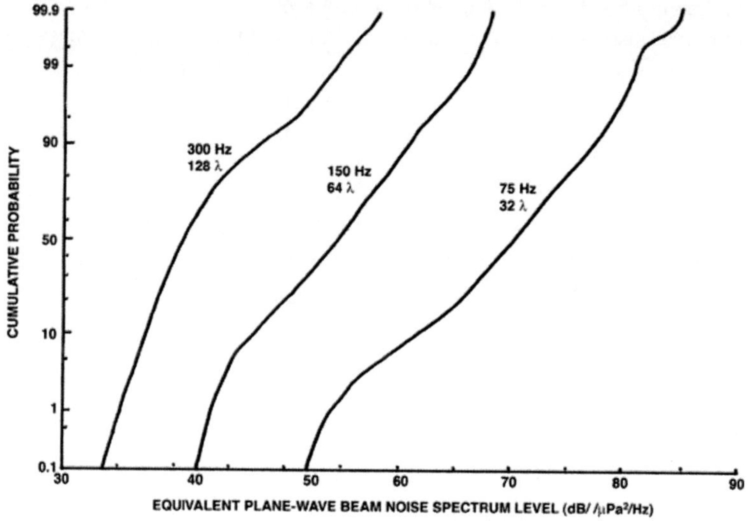

A. The Ionian Sea Cumulative Distribution Function

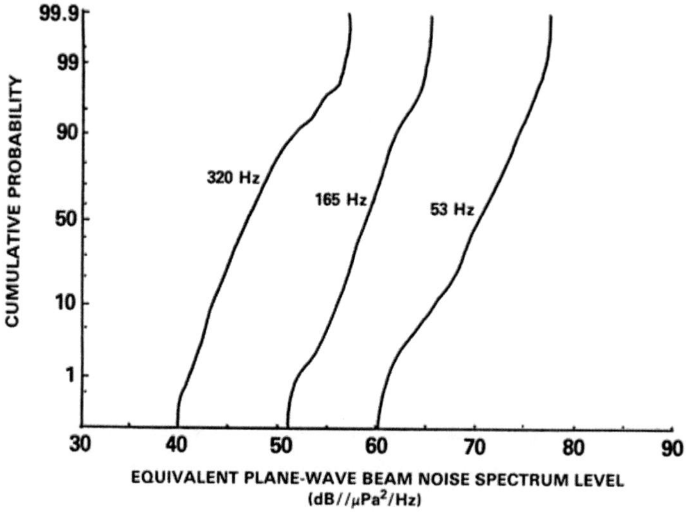

B. The Levantine Sea Cumulative Distribution Function

Fig. 4.18 Cumulative distribution functions for array beam noise levels for the Ionian Sea and the Levantine Sea. Each beam output was normalized to the plane wave response of the system. Each beam noise level was determined from a 0.18-Hz bandwidth, an 8-s coherent integration time, and an incoherent average over 6 min with the background noise, the lighter areas. The dynamics of the two-component noise field, shipping and environmental, are obvious.

Cumulative distribution functions were computed for various measurement parameters to quantify the statistical characteristics of beam noise. The cumulative distribution functions for three frequency bands with an integration time of 8 s showed insignificant differences. The beam noise autocorrelation function, using the complex FFT beam output, was used to determine the average decorrelation time, taken as the zero crossing, of 10 min. Cumulative distribution functions for the broadside to aft end-fire beams based on 6-min averages from the Levantine Sea and the Ionian Sea were found to have substantial differences. The Levantine site was close to major shipping lanes, whereas the Ionian site had a more uniform distribution of ships. These cumulative distribution functions are decidedly not log normal, and are consistent with the dynamical distribution of ships and the relative response of the beams and side lobes to their radiated noise.

Vertical Noise Directionality

In a deep cylindrical ocean basin, the sound velocity maximum near the surface decreases as the depth increases to a minimum, and then increases as the depth increases owing to the increase in pressure. This sound speed variation produces a sound channel between the surface maximum and the critical depth at which the speed of sound has reached the surface value. Figure 4.19 shows this variation for a representative basin with a continental slope (3–6°) and shelf.

Vertical array measurements such as those by Fox (1964) and Axelrod et al. (1965) showed at higher frequencies the deepwater noise directionality was primarily from surface-generated noise whereas at lower frequencies noise along the horizontal was observed Fig. 4.20. Cron et al. (1962, 1964) explained these measurements with space-time correlation functions derived from volume sources (isotropic noise) and surface sources (anisotropic noise) and showed the vertical directionality at higher frequency to be proportional to $\cos(\theta)^m$, where θ is the depression angle and $m = 1, 2, \ldots$. This type of analysis was extended by Cox (1973) with the utilization of spherical harmonics to derive the general space-time correlation functions that facilitated the estimation of pairwise covariance properties. His results agreed favorability with a variety of measurements and provided a method of determining the optimum spacing of array hydrophones.

Sound radiated from the surface or near the surface was consistent with dipole-like directional characteristics and produced a characteristic directional pattern on a vertical array. Talham (1964) derived an expression for the vertical directionality from surface noise sources incorporating the effects of absorption, refraction, and reflection as a function of receiver depth. This type of analysis shows downward, higher-angle, directed energy is absorbed by multiple interactions with the boundaries and results for a flat ocean basin in the arrivals being peaked at the SOFAR channel limiting angles (10–15° off the horizontal). However, as shown in

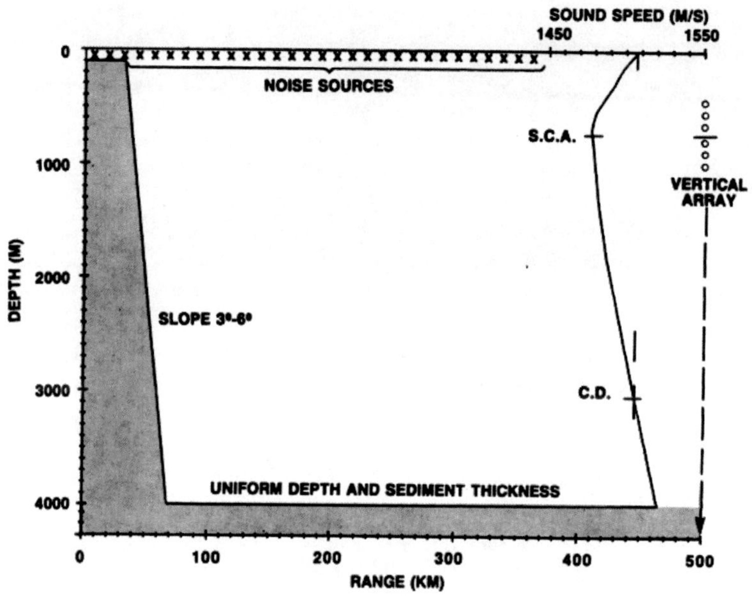

Fig. 4.19 The typical basin cross section with a deep water sound speed profile showing the sound channel axis and the critical depth

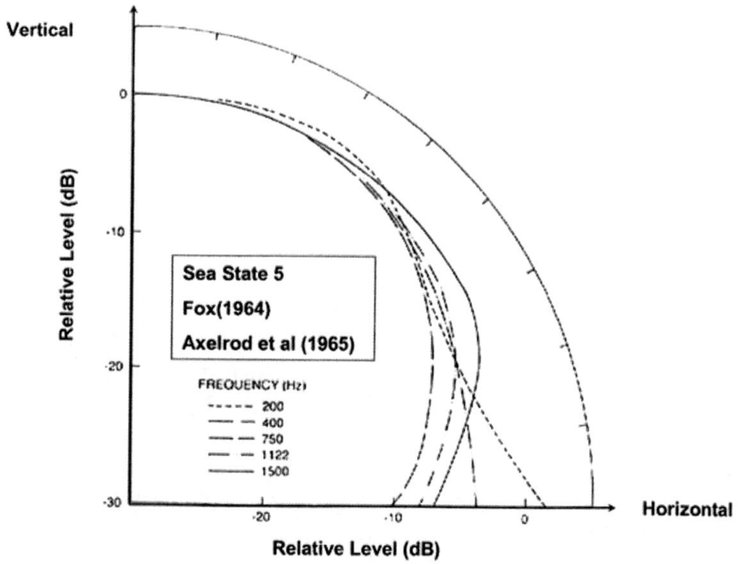

Fig. 4.20 The vertical directionality of noise, relative level (dB), versus elevation angle for frequencies between 200 and 1,000 Hz

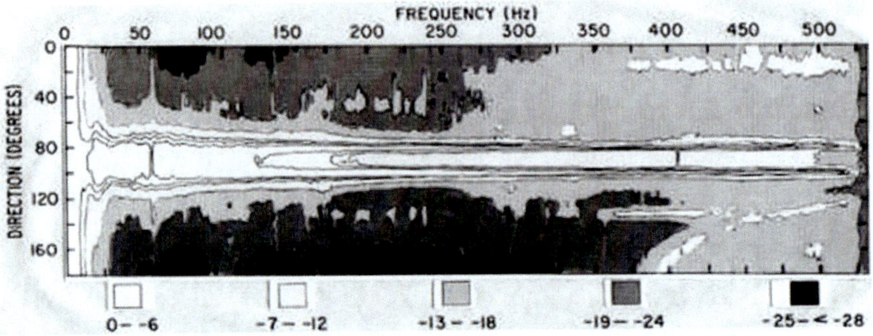

Fig. 4.21 The vertical arrival structure observed on an array in the sound channel as a function of vertical angle and frequency (Carey and Wagstaff (1986, Fig. 4)

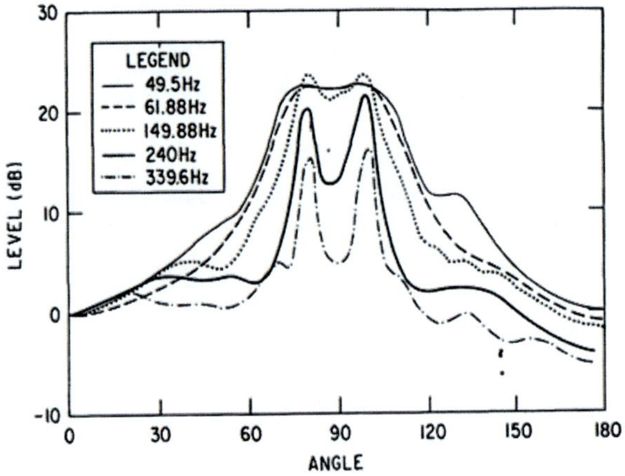

Fig. 4.22 Noise level versus vertical angle (0° is up, low sea state) taken from Fig. 4.18 at specific frequencies. The broad maximum along the horizontal at 49.5, 61.88, and 149.88 Hz is compared with the peaked distributions at 240 and 339 Hz

Figs. 4.21 and 4.22, measurements of the vertical noise directionality with an array in the sound channel (Fig. 4.19) have at the lower frequencies a broad pedestal centered on the horizontal direction, whereas at higher frequencies the pattern is peaked at the sound SOFAR angles. For an array at a different depth, the directionality, of course, will change with the arrival structure. For example, if the array were at the critical depth, a different directional pattern would be observed at all frequencies. An approximate ray angle analysis of the vertical directionality provides insight as to the variation of vertical directionality with depth. Sources of sound near the surface of the sea are dipolelike; consequently, the radiation pattern is directed downward and especially at higher frequencies is attenuated by multiple bottom interactions, thermocline variations in the sound speed, and rough surfaces interactions.

The measured results shown in Figs. 4.21 and 4.22 (Carey and Wagstaff 1986) were obtained by Anderson et al. (1972) with the maximum likelihood method of Edelblute et al. (1966) using a 26-element array in the Sargasso Sea south of

Bermuda. The array center was at a depth of 236 m in the sound channel with an axis depth of 1 km and was processed with a frequency resolution of 1.4 Hz.

Figure 4.21 shows the low-frequency plateau that begins to reduce in level at frequencies higher than 150 Hz to a "notch" (10 dB) along the horizontal (also see Fig. 4.22). The noise intensity at 150 Hz has a maximum at $90 \pm 9.5^{\circ}$, compared with $90 \pm 14^{\circ}$ if the array had been centered on the sound channel axis. A reference sound source at some distance on the axis is observed near 400 Hz. The results also show evidence of aliasing at frequencies greater than 400 Hz.

Since surface spectra are tonal in the low-frequency range, one would expect more of a variation of the level with frequency along the horizontal. The smooth variation of this pedestal along the horizontal with frequency indicates that broadband, wind-induced sound in addition to ship-radiated noise may be important. Any source over the basin shelves, slopes, and seamounts will reflect sound into the channel, as the calculations using monopole sources a small distance below the surface show in Fig. 4.23. The calculations ignore local surface noise sources in the array vicinity but have a distribution in range of random noise sources with source levels based on a 10-knot wind speed. The calculated mid-basin omnidirectional noise levels of 65 dB at 50 Hz, 53 dB at 200 Hz, and 45 dB at 400 Hz were consistent with measured levels when only wind-induced noise is the source of sound. The 1974 measurements of Garabed and Finkelman (2005) and later measurements of Wales and Diachok (1981) in the Northwest Atlantic show this vertical directionality of the noise.

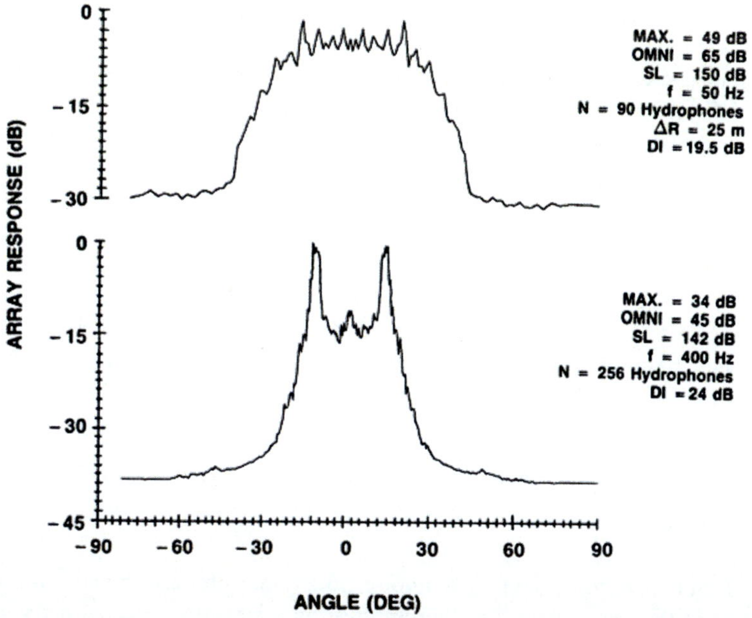

Fig. 4.23 Relative noise levels versus angle (+90° is up) for 50 and 400 Hz for the geometry and noise source distribution shown in Fig. 4.19

Since range-dependent sound velocity profiles, boundary reflections, and sound channel scattering couple energy from higher to lower angles and produce frequency-dependent losses, the combination of these effects was thought to be the primary cause of the vertical directivity characteristic. Although all are important, it appears that the first-order explanation is the downslope conversion of noise sources over the slopes and seamounts. The range-dependent environment in Fig. 4.19 was employed to numerically investigate these effects and a sample result is shown in Fig. 4.23. The sources were placed at a reference depth to ensure there were dipole or doublet characteristics and a scaled source level per unit area consistent with measurements of wind noise. The low frequency (50-Hz) sound is reflected from the slope with little loss and has a broad pedestal centered along the horizontal, whereas the higher frequency (400 Hz) with large loss produces a minimum along the horizontal. Using the same numerical method, the decrease in the sound channel toward the northern latitudes was examined and found to produce a smaller effect. The depth of the noise notch is shallow at higher frequencies owing to sound channel scattering from the thermocline and bathymetry changes. An interesting characteristic of Fig. 4.21 is the smooth frequency characteristic along the horizontal for frequencies less than 140 Hz.

Coincident measurements of the horizontal and vertical noise characteristics showed that when the horizontal directionality was successfully modeled only using surface-ship-radiated noise, the level of the horizontally directed noise and the subsequent pedestal in the vertical noise level distribution were underpredicted.

The experiments conducted by Wagstaff (2005) and Anderson (1979) featured the measurement of ship density and traffic using airborne observations, the measurement of horizontal directionality with high-resolution arrays, and the measurement of vertical noise directionality with an array that spanned the sound channel. The comparison of the measurements of the persistent directionality and the calculated directionality using the range-dependent bathymetry, sound speed structure and observed density of ships is shown in Fig. 4.24. Moderate agreement was found and is rather remarkable when one considers the estimated bottom reflectivity and source levels of the shipping. The comparison of the calculated and measured vertical directionality is shown in Fig. 4.25. Observe that the noise levels along the horizontal are higher for the measurements. This difference in the relative levels as a function of the angle from the horizontal indicates the calculations do not include all the noise sources over the shelf, such as wind-generated noise. This frequency-dependent characteristic also depends on the not well known slope reflectivity.

Nevertheless the downslope enhancement, megaphone effect, is an important factor.

Rain Noise

Many factors determine the radiated noise from rain. Certainly, the rate of rainfall, the state of the sea surface, and the presence of a subsurface bubble layer can be important with respect to the level and directional characteristic. The Wenz curve

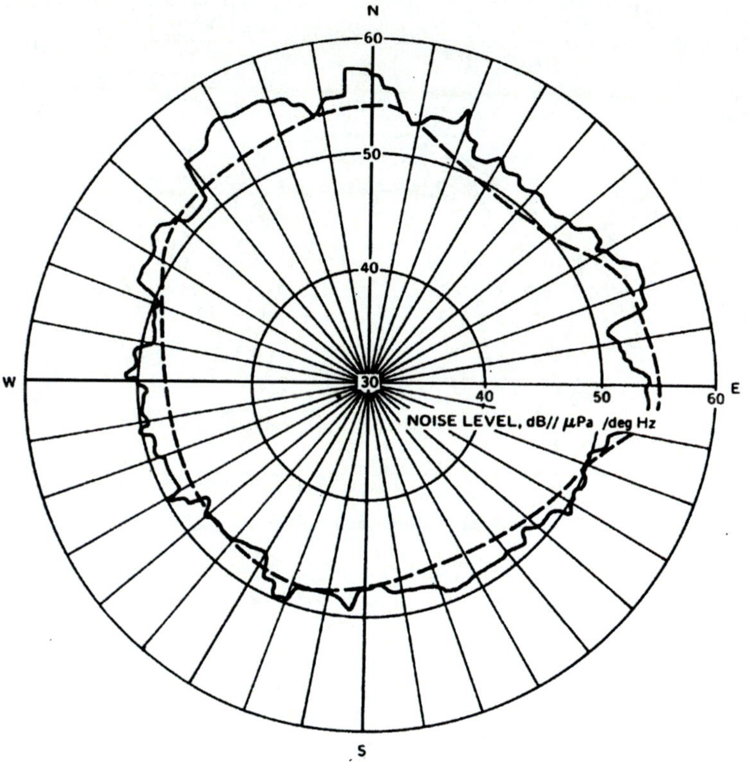

Fig. 4.24 Measured horizontal directionality in the Northeast Pacific Ocean (*solid curve*) compared with calculations based on range-dependent bathymetry, range-dependent sound velocity profiles, and measured density of ships. (Wagstaff 2005) and (Urick 1984)

presented in Fig. 4.2 shows the results from Heindsman et al. (1955) and represents the maximum noise levels measured during the passage of hurricane Edna. The sea state was 4 and the wind ranged from 20 to 40 knots. Heindsman et al. reported rainstorm measurements with a rainfall rate of 9 mm/h with a flat spectral characteristic from 0.1 to 20 kHz and a spectrum level of 80 $dB\ re\ ((1\mu P_a)^2/Hz)$, equivalent to the levels measured by Knudsen et al. under similar conditions. The striking feature of the spectrum under the condition of heavy rains is its white spectral character. This characteristic was not generally observed for lighter rainfall rates, which produce a spectral peak in the 10–15-kHz region primarily due to oscillation of entrained bubbles. Fig. 4.26 shows the impact of a drop and an entrained bubble oscillation.

Franz (1959) (also see Fitzpatrick and Strasberg (1956, Fig. 18, p. 264)) performed a remarkable series of experiments of drops impinging on a liquid surface. His experiments were guided by the theoretical expectation based on a multipole expansion of a sound source at the free surface and the result that the radiated pressure would have the dipole radiation form:

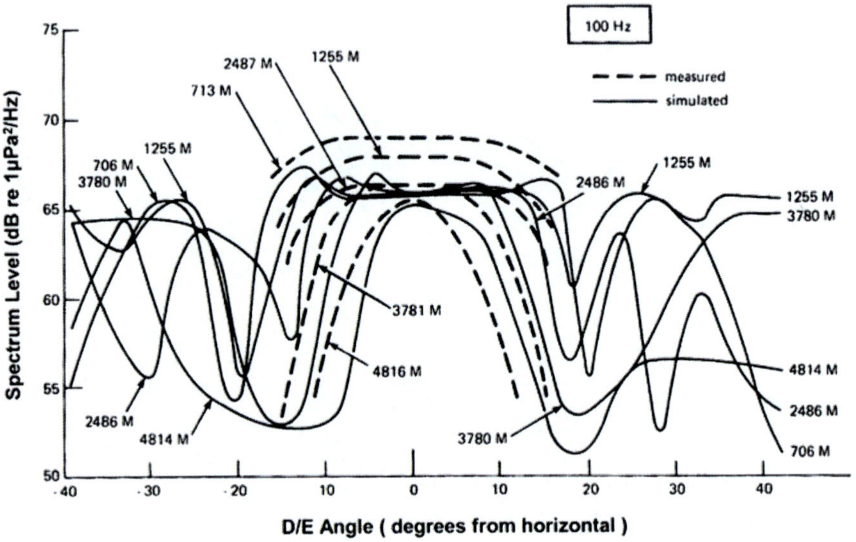

Fig. 4.25 Measured vertical noise directionality compared with calculations for the same case shown in Fig. 4.24 showing underprediction of the levels along the horizontal

$$p_s(r, \theta, t) = (\rho\, u^3 L_d \cos(\theta)/rc) Z((u/L_d)(t - r/c)),\ L_d = 2a., \tag{35}$$

where ρ is the density, u the impact speed, θ the polar angle, a the drop radius, and r the radial distance to the point of observation. Figure 4.26 shows one of the experimental sequences that identified two sources of sound, the impact and entrained bubble oscillation. Although a fairly white spectrum was observed, was the drop impact the dominant mechanism? Franz always observed the impact but not the bubble oscillation

The impact waveform and measured spectrum, shown in Fig. 4.27, interpreted with the above expression for the radiated pressure define the radiated sound of the impact. However, comparisons with measurements (Fig. 4.28) reveal that Franz's theoretical expectations produce levels less than observed by an order of 10 dB.

Guo and Ffowcs Williams (1991) reexamined the drop impact and determined that the cause of the radiated sound was the rapid momentum exchange between the drop and the water body. They applied the Kirchhoff integral theorem to develop an expression for the radiated pressure in terms of the vertical velocity at the pressure-release surface:

$$P(r, t) = (\rho_o/2\pi) \cdot \partial/\partial t \int_s \{u_3(y_x, \tau)/r\} dS_y. \tag{36}$$

This is equivalent to one of the integral expressions (Eq. 21) developed in Chapter 3. They further showed that the supersonic contact circle was the cause of the short-duration compressive waves. As the contact circle becomes subsonic,

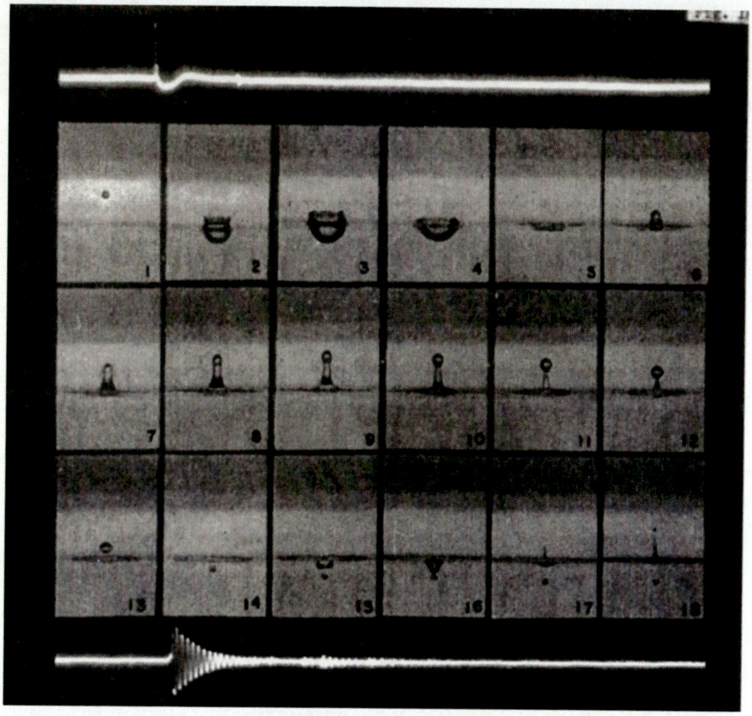

Fig. 4.26 The sequence of events of a liquid drop impinging on a water surface is shown by a series of photographs every 13 ms. Also shown is the corresponding oscilloscope trace of the impact (frames 1 and 2), the dampened sinusoid due to the bubble (frames 13 and 14), and the subsurface bubble frames (14–18). The radius of the droplet was 0.24 cm, the mass was 56 g, and the droplet speed was 350 cm/s. (Fitzpatrick and Strasberg 1956, Fig. 18)

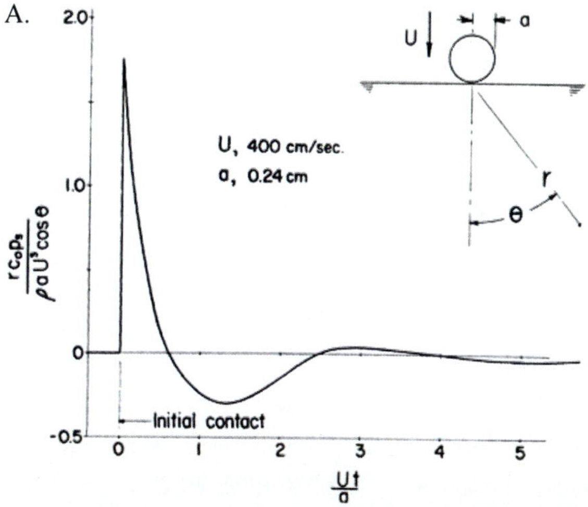

Fig. 4.27 The impact waveform (**a**) traced from an oscillograph record and the spectrum (**b**) are shown to characterize the sound produced by the vertical impact of water droplets

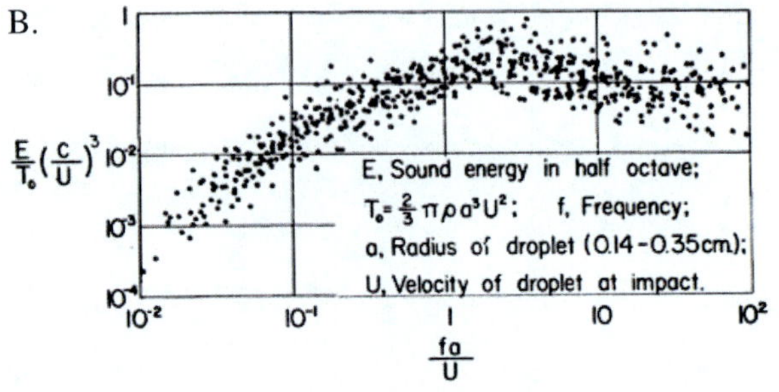

Fig. 4.27 (continued)

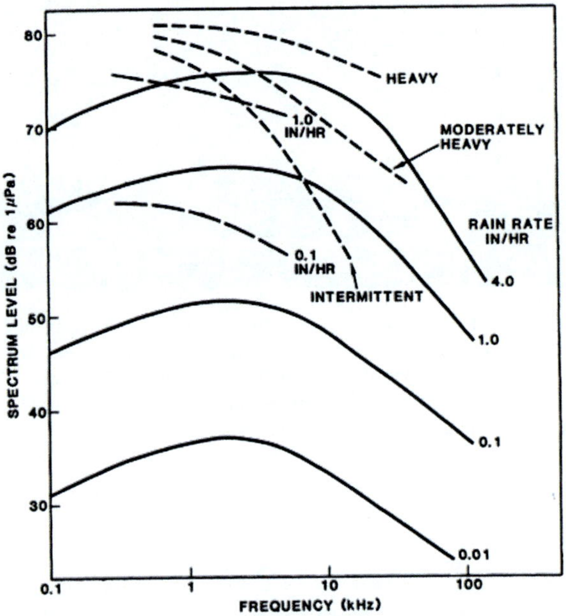

Fig. 4.28 Measurements of the rainfall spectra by Heindsman et al. (1955) (*short dashes*) and Bom (1969) are shown in comparison with the theoretical estimates by Franz

the horizontally propagating wave has particle velocities greater than those of the contact circle expansion and, thus, disturbs the fluid ahead of the contact circle, reducing the relative velocity and, subsequently, the generation of additional sound. The result is a short-duration transient and slowly decaying tail. Since the initial impact is essentially a "water hammer," the initial pressure amplitude, p, is proportional to $\rho_o Uc$; the subsequent three-dimensional propagation of the pulse reduces this amplitude. The radiated energy of the impact determines for a monodispersed distribution of drops the far-field intensity for a given rain rate:

$$R = (4\pi a^3/3)N, \quad N[\#/s]; \tag{37}$$

to be

$$I = (1/8)\rho_o U^2 \pi\, a^3 M^3 N = (3/32)\rho_o U^2 M^3 R. \tag{38}$$

If only the impact sound is important, the intensity of sound in moderate–heavy rain should be related linearly to the rainfall rate R. However, many factors change as the sea state increases, such as the surface roughening, waves breaking, and near-surface bubble layers.

The question addressed by Pumphrey and Crum (1989) and later by Pumphrey and Elmore (1990) was when are bubbles produced? Their answer, shown in Fig. 4.29, was that bubbles are produced by (1) irregular entrainment of bubbles caused by complex splashes, (2) regular entrainment caused by the crater resulting from drop impact and the formation of a smaller individual bubble, (3) entrainment of larger bubbles formed from the volume of the crater, and (4) Mesler entrainment of many small bubbles formed between capillary wave crests and at the instant of impact.

The importance of bubbles as sources of sound has been well established and if bubbles are produced by every impact, then their contribution to rain noise is beyond question. The simplified analysis of Franz ignored two important nondimensional

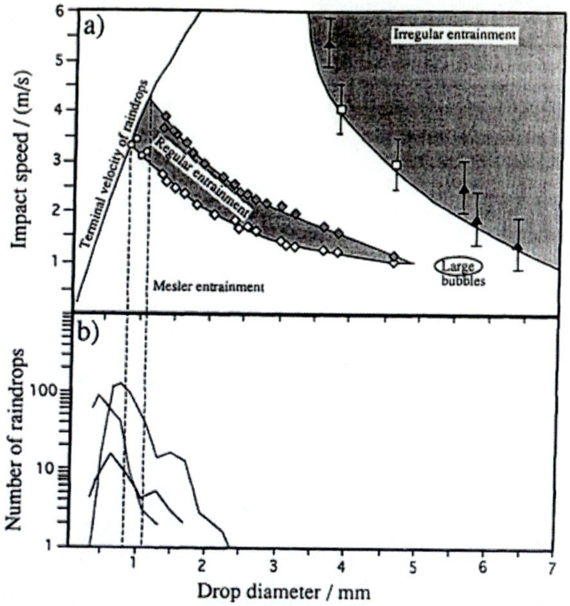

Fig. 4.29 **a** Rain drop impact speed versus the drop diameter for regions of irregular entrainment, regular entrainment, entrainment of larger bubbles, and Mesler entrainment. **b** The size distribution of rain drops, with the abscissa coincident with **a**

ratios: the Froude number, $Fr = U^2/L_d g$, and the Weber number, $We = \rho\, U^2 L_d/T$. These nondimensional ratios have a key bearing on bubble entrainment. Pumphrey and Elmore (1990) found that the boundary of the regular entrainment region (Fig. 4.29) corresponds to a relationship between these numbers as $Fr{=}A\, exp(BWe)$. In this expression A and B are constants, one pair for the highest drop speed and another for the lowest drop speeds that cause entrainment. Although this work is interesting, the fundamental characterization of the drop-entrainment process is not fully understood. Also shown in Fig. 4.29 is the bubble size range of 0.8–1.1 mm corresponding to normal-incidence drops at their terminal velocity impacting on the surface.

The radiation of a bubble near the surface can be described by (see Appendix G)

$$\begin{aligned} p_D &= ikD\, cos(\theta)(1 + i/kr)\, exp(-\beta(t - r/c))\, exp(-i\omega_o t + ikr) \\ &\rightarrow ikD\, cos(\theta)\, exp(-\beta(t - r/c))\, \exp(-i\omega_o t + ikr)\ ; \end{aligned} \tag{39}$$

this equation describes a progressive dampened sine wave with frequency $f_o = 2\pi\omega_o$. The natural frequency of the bubble is given by the equation of Minnaert (1933) (see Appendix E),

$$f_o = (1/a_o)\sqrt{3\gamma P_o/\rho}. \tag{40}$$

The spectral peak of the radiated noise from small bubbles is determined by the bubble size distribution and the dampening constant. This phenomenon reported by Nystuen for light rains has been widely observed. Notice as the wind speed increases in Fig. 4.30, the spectrum level decreases owing to the suppression of the regular entrainment of bubbles shown in Fig. 4.29

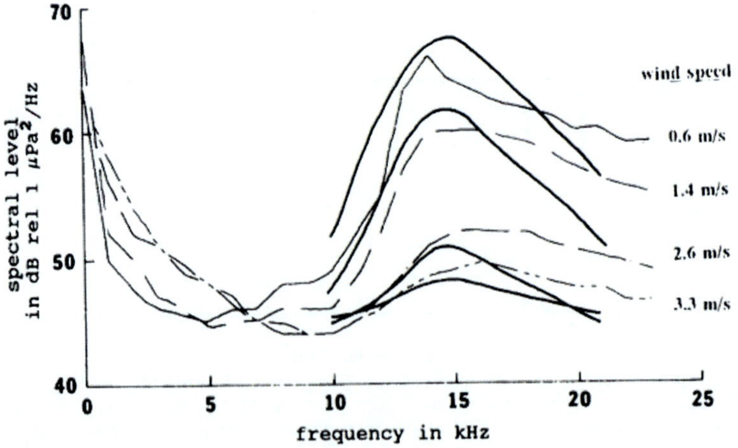

Fig. 4.30 The spectrum of light rain for various wind speeds (Nystuen 1993)

The spectral peak in the 12–15-kHz range corresponds to the regular entrainment of bubbles by drops at their terminal velocity. Also shown are Nystuen's prediction of the levels based on the drop size distributions of Pumphrey and Elmore (1990).

The rainfall sound intensity and spectrum characteristic can be described largely by the cumulative effects of many raindrop impacts that do not produce bubbles. However, when the rain is heavy, the broad spectral peak observed by Heindsman et al. in the region below 3 kHz requires another source such as the irregular entrainment of larger bubbles.

Arctic Ambient Noise

The Arctic Ocean can be considered to consist of four relatively distinct regions: the central Arctic, which is covered with pack ice; the coastal regions, which have shore-fast ice in the winter and a mixture during the summer; the marginal ice zone, which represents the progression from pack ice to an ice floe region; and, finally, the open waters, which are ice-free and adjacent to the marginal ice zone.

The central characteristic of the Arctic sound channel is the presence of ice and, consequently, it represents a unique noise environment. Noise is produced by the interaction of the pack ice and floes with the environmental forces such as the wind, rate of temperature change, radiation cooling, insolation, and mesoscale oceanic currents. These sources of sound are temporally variable as well as seasonal in nature. Because of this variability, the noise field can be up to 20 dB less than the level measured in the open ocean for a Knudsen sea state zero (Beaufort 1 in Fig. 2.2, Table 2.1) and up to levels corresponding to sea state 4 (Beaufort 5 in Fig. 2.2, Table 2.1). At low frequency, naturally occurring quakes, microseisms, ice cover standing and propagating surface waves, and water sediment boundary waves can all contribute to the noise field.

In addition to noise that originates from the ice, there is a rich source of both narrow-band and broadband noise from the soniferous marine fauna. During the summer, sounds such as long gliding whistles, short chirps, braying, groans, and grunts have been observed. However, biological noises in the Arctic are not dominant since the levels and duration are limited. On the other hand, during the summer in the Antarctic's McMurdo Sound, noise from seals and humpback whales dominates the spectrum from 200 to 800 Hz.

Finally, the Arctic noise field consists of man-made sounds from explosions, air guns, industrial activity, and icebreakers. These noise sources will not be treated even though their contributions to the ambient noise field can be at times significant.

Arctic Ambient Noise over the 10 Hz to 1 kHz Band

The interaction of the ice cover with the air and water boundary layer is the primary source of noise production. Since the ice covering has variable characteristics, such as brittleness, snow cover, thickness, surface roughness, and ridging, variability in noise production is to be expected. Furthermore, the general forcing functions such as rate of temperature change, wind speed, oceanic currents, and tidal variations as well as the interaction with the ice are very complex. It is doubtful whether one could predict quantitatively the observed noise field; rather, it is the goal in this section to provide a qualitative understanding of the dominant observed characteristics.

Figure 4.31 shows the ambient noise results of Mellen and Marsh (1965). Curve A represents a quiet period over the frequency range of 10–1,000 Hz. The broad spectral peak observed around 30 Hz was thought to be due to distant sources. Curve B is representative of a cold period with a sudden change in temperature that produced thermal-stress-induced ice cracking. Curve C represents the highest noise levels observed.

These early observations of Arctic ambient noise are consistent with the work of Greene and Buck (1964), Milne and Ganton (1964), Milne (1965), and Ganton and Milne (1965). The basic cause of higher-frequency noise was considered in each work to be thermal-stress-induced cracking of the ice during cold periods with rapid decreases in temperature. By "cracking," one means a tensile opening and shear motion either toward the direction of or across the crack producing a series of slip–stick sounds. Surface cracks can be caused by radiation cooling and require a snow-free brittle ice. This effect is often observed in pack ice and shore-fast spring and winter ice. The ensemble of stick–slip emissions results in a short burst

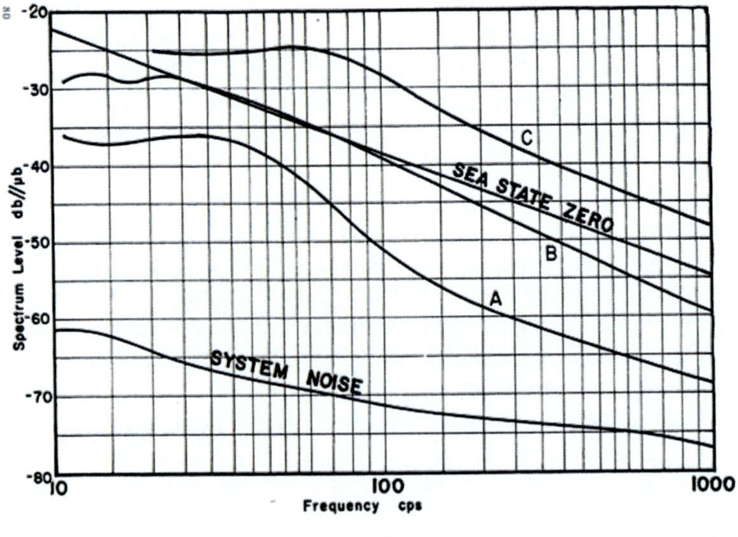

Fig. 4.31 Summary of ambient noise measurements from two ice islands made by Mellen and Marsh (1965). The ordinate is labeled "Spectrum level dB//μb." To convert to spectrum level dB re (μPa)^2/Hz simply add 100 to the ordinate. *Curve A* represents a typical quiet period. *Curve B* is representative of cold weather and a sudden decrease in air temperature. The rapid decline in air temperature causes the ice to become stressed and to crack. *Curve C* represents the highest levels observed

of sounds. These spiky and random slip–stick emissions, when not closely spaced in time, are decidedly non-Gaussian. However, when the number of transients per unit time produces a random waveform without identifiable individual events, the process becomes Gaussian.

Milne (1967) observed for the higher frequencies that two prime causes of noise were evident, wind speed and temperature change. To examine these causes he conducted experiments under similar ice conditions during the colder cooling periods with no solar radiation and during warming periods with variable winds, as shown in Fig. 4.32

During cooling trends, clear conditions and low (about $0\,m/s$) wind speed conditions, Milne observed te spectrum levels shown in Fig. 4.32a. Figure 4.32a shows the results for periods during which there was no wind but cooling, as shown by curves C, D, and E. Curve "A," labeled "residual impulsive noise," represents a limiting condition and was obtained under a period of warming with low (about $0\,m/s$) wind speed with levels 14 dB less than a sea state "0" condition. Also shown is a reference curve for noise levels with a 10-knot wind speed. The broad maximum

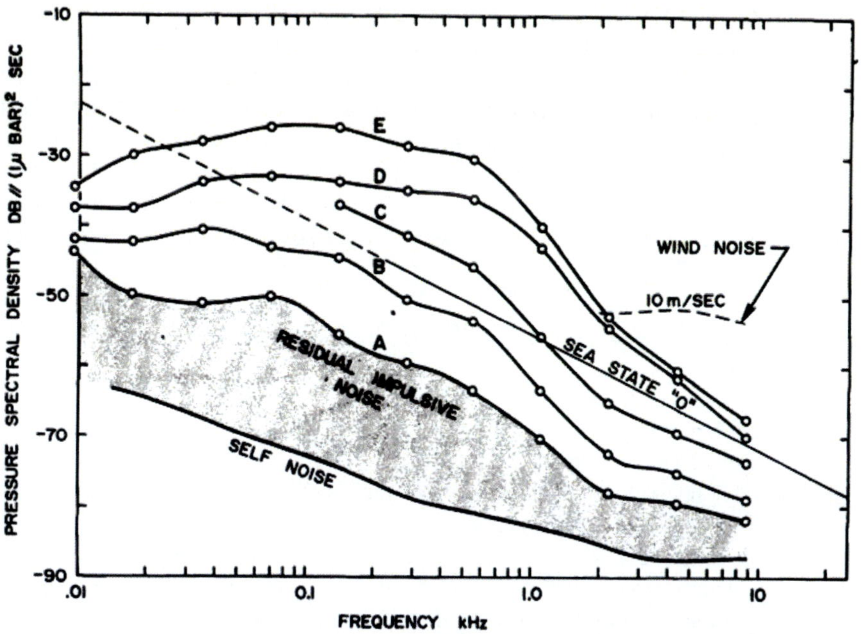

Fig. 4.32 Under ice pressure spectrum levels versus frequency during cooling with low (about $0\,m/s$) wind speed and under warming periods with increasing wind speed. The ordinate can be converted to $dB\ re\ (\mu\ Pa)^2/Hz$ by simply adding 100. (Milne 1967, Figs. 7.34 and 7.36). **a** Pressure spectrum level versus frequency of under ice noise measured during an air cooling period with low (about $0\,m/s$) wind speed illustrating ice cracking noise. **b** Pressure spectrum level versus frequency of under ice noise measured during an air warming period for various wind speeds

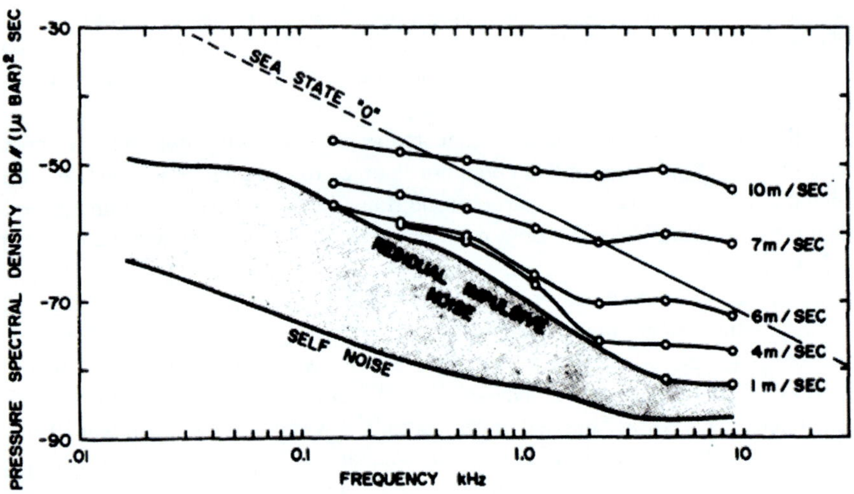

Fig. 4.32 (continued)

is in the 100–600-Hz range. Milne attributed the spectral characteristic of the noise during the cooling period to ice cracking, a large collection of impulsive events.

Shown in Fig. 4.32b is the case for air warming periods with variable wind speeds. Unlike the impulsive ice cracking noise, the observed wind speed noise had a whitened spectrum and a Gaussian amplitude distribution. This figure shows the noise levels for various wind speeds (measured at 5 m above the ice surface) as a function of frequency. The increase of the noise intensity is observed to be approximately proportional to the 5th power of the wind speed. The empirical equation developed by Milne is

$$<p^2> = v^{5.3} \times (1.6\,10^{-1}),\ [(\mu P_a)^2/Hz], \tag{41}$$

where the wind speed v is in meters per second.

The turbulent boundary layer can produce ice surface pressure fluctuation owing to the unevenness of the surface and, thus, can generate noise. However, this turbulent boundary layer can also cause snow pelting. That is, wind flow over the ice can produce miniature eddies of small snow and ice crystals that collapse in the lee to produce blowing snow, sounding very much as if the ice surface were being blasted with many small particles (Mellen and Marsh 1965). These sounds were found to have a Gaussian amplitude distribution. Milne and Ganton (1964) also observed that snow pellets can be lifted from the surface by flow eddies and subsequently impacted on the surface. For wind speeds greater than 4 m/s and for frequencies greater than 2 kHz, this wind speed noise can dominate. As one proceeds to lower frequencies, longer-range or distantly generated events are observed. Wind-induced noise should be seasonal and primarily a winter phenomenon. During the summer months, stable temperature and wind conditions result in quiet ambient noise levels since broken ice is a poor noise maker as the slow jostling of floes does not much produce significant noise.

Low-Frequency Arctic Ambient Noise

An interesting phenomenon has been observed in the marginal ice zone: the dramatic increase in noise as one approaches the ice edge (Diachok and Winokur 1974). The results from Yang et al. (1987) shown in Fig. 4.33 were obtained in July over a 2 h period under sea state 2 conditions by use of air-dropped sonobuoys into the open water and polynas, which are water openings in the ice field. The nominal spacing was 28 km and the depth was 30 m. Measurements were made over a frequency range from 100 to 1,000 Hz. A sharp increase in noise at the compact ice edge was observed across the whole 100–1,000-Hz band and Fig. 4.33 shows the effect for a one-third-octave band centered at 315 Hz. The compact ice edge corresponds to ice concentration changes of one eighth to seven eighths of the area covered over a range of 1–2 km.

Also shown in Fig. 4.33 are the results from measurements made during April in the vicinity of a diffuse ice–water boundary corresponding to ice concentration changes of one eighth to seven eighths of the area covered over a range of 100 km. The spatial variation of the observed noise levels in the band centered on 315 Hz was

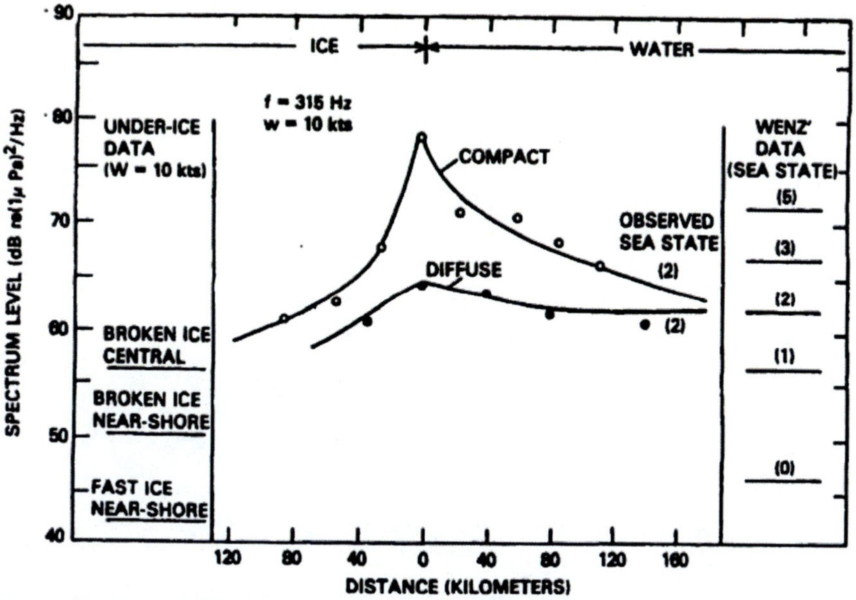

Fig. 4.33 Ambient noise results obtained in the Greenland Sea near the boundary of compact and diffuse ice floes and the open ocean. Also shown are the average noise levels under the ice-free and ice-covered regions. (Yang et al. 1987)

found to be more spatially uniform with a less pronounced increase in noise level near the edge. These results are consistent with those from other Arctic marginal ice zone investigations (Urick 1984, pp. 8-12 –8-14).

The result from all these investigations was that the ice edge acts as a distributor of sound producing higher noise levels on the open-ocean side of the zone and lower levels under the ice, with a broad maximum along the ice edge. This distribution of sound radiators along the edge was not in general uniform. Directional arrays towed at depth in the open-ocean areas parallel to the ice edge showed that regions of intense noise activity were observed and persisted for days.

The bumping and crashing of ice floes produced intermittent strong transients in the 100–1,000-Hz frequency band. The ice crush spectra were similar to those of thermal cracking and ice breaking. Since ice edge noise increased with wind speed and the density of floes, the combined role of wind, waves, and currents cannot be understated.

At very low frequencies, standing waves and flexural waves can occur on individual ice floes that may produce low-level pressure oscillation independent of depth at twice the frequency of gravity waves and the bobbing motion of ice floes:

$$Period\ T = 2(h\rho/g\rho_w)^{1/2},\ h = ice\ thickness = 2 - 3m,\ f_b = 0.35 - .7\,Hz \quad (42)$$

Figure 4.34 shows an interesting composite of low-frequency ambient noise results compiled by Dyer (1983). Each portion of the spectrum was obtained at

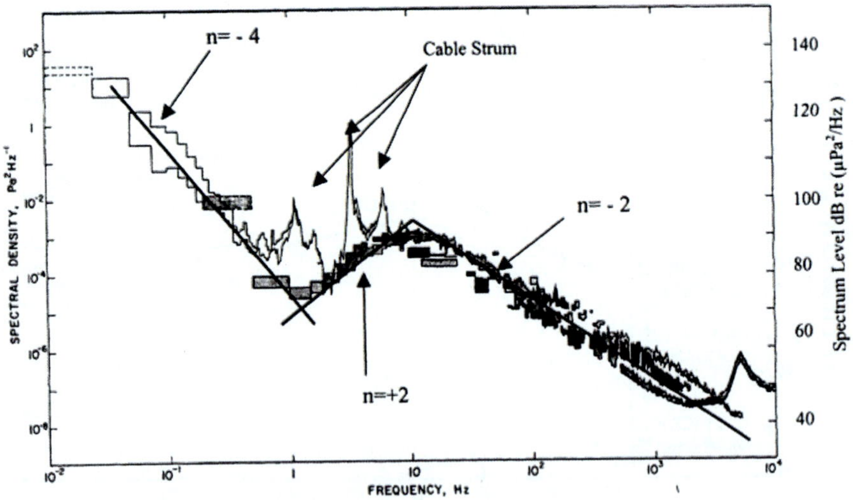

Fig. 4.34 Composite noise results from multiple sample periods under pack ice (83° N, 20° E, 1982) are shown to illustrate low-frequency noise characteristics. Also shown are the frequency dependencies of the spectral levels, $(f/f_{ref})^n$, and identified spectral features due to strum. [Adapted from Dyer (1983, Fig. 7, p. 24)]

different periods and placed in the composite to enable discussion of the basic noise characteristic and causative physical source mechanisms.

In the low frequency of the spectrum, 0.01–0.5 Hz, the spectrum decreases with approximately the fourth power of the frequency, n = −4, and Dyer considered this noise to be the result of hydrodynamic flow, subsurface oceanic currents, around the hydrophone and cable. He identified the spectral features caused by the flow-induced strum of the suspension cable coupled to the hydrophone.

A striking spectral feature was the broad low-frequency spectral peak near 10 Hz between 1 and 100 Hz, with frequency to the n = +2 power prior to the peak followed by frequency to the n = −2 power. This characteristic was found to be a regular feature of pack-ice noise. The region between 150 and 5,000 Hz was similar to the results of Milne discussed previously due to thermal cracking of the ice in periods of rapid cooling. Another peak was observed near 5,000 Hz during periods of high winds, most probably due to snow pelting.

Shepard (1979) concluded that the cause of the low-frequency spectral feature was a superposition of a short burst of sound; see Fig. 4.35. Even though the individual event may have a narrow steeper-sloped spectrum, the superposition of many such impulses would result in the spectral characteristic observed.

The "Arctic song" of Dyer, the broadly peaked spectrum from 1 to 100 Hz with a central peak in the 10-Hz band, was attributed to cracking and breaking of the pack ice. Ice, like many crystalline and regular granular materials, when under stress can crack and break. The acoustic emissions from such events are caused by stick and slip action of the breaking or cracking motion. Generally speaking, the emissions from many such events yield a lower-frequency, longer-duration acoustic emission. This phenomenon occurs from the seismic to ultrasonic scales. The resulting waves generated then propagate and are filtered by the response of the transmission path. Dyer (1984, pp. 13–15) provided an analytical treatment of this process by assuming that the shear displacement radiated from a small slip region in an infinite medium can be represented by a Gaussian proportional to the area-averaged slip displacement and the derivative of the time-dependent slip moment, $\dot{M}(t - r/c_s)$, with a retarded potential. The pressure radiated into the water requires the treatment of the ice–water interface. The traction across this interface introduces another time derivative of this moment due to the retarded time. Finally, another time derivative results from the ice–air boundary. Dyer concluded the pressure observed in the water is

$$P(r, t) = H(r)\dddot{M}(t - r/c_s) \tag{43}$$

where H is the total transfer function and c_s is the shear wave speed. This treatment, however, yielded the result shown in Fig. 4.35 with spectral slopes different from those observed.

Milne (1965) performed analysis of acoustic emissions from the ice. He assumed that the individual events were dampened sinusoids with random amplitudes, frequencies, locations, and time of occurrences. He derived the spectral characteristic

A.) Amplitude versus time

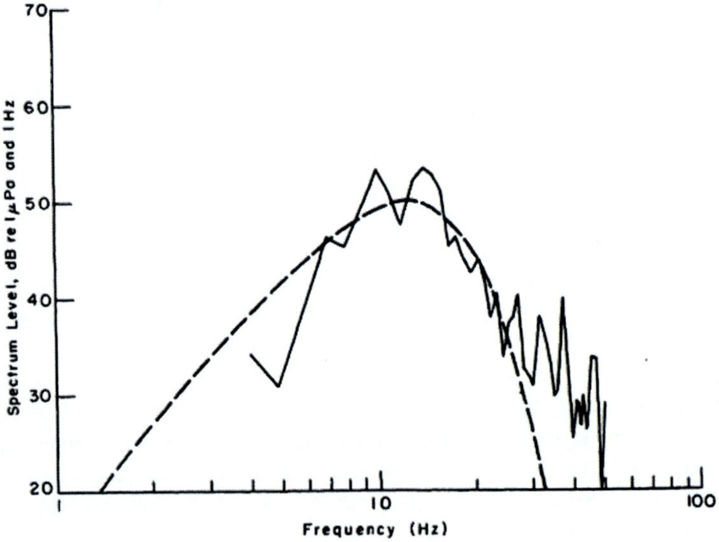

B.) Transient capture spectra of one impulsive event.

Fig. 4.35 The spectral characteristic of impulsive pressure pulses caused by stick–slip phenomena at the low frequencies that are generated by composite moments and stress due to environmental factors (Dyer 1983, Fig. 10)

using characteristic functions to show that at higher frequencies the spectral decline goes as $n = -2$. A similar analysis could be done on a collection of Gaussian pulses.

Since the measured acoustic pressure should be proportional to the third derivative of the slip displacement, noise is created when the ice fractures. The fracture or breaking of the ice is caused by the net stress and moments on the pack-ice sheet. Makris and Dyer (1986) chose as variables the composite ice stress, S, and composite ice moment, M, determined from the wind, current, cooling, and drift measurements. At low frequency, they concluded that only the wind current and drift were important. This conclusion was based on the temporal variation of the smoothed root mean square pressure in the 10–20-Hz band and its correlation with the temporal variations of the interrelated environmental variables, the effective stress S and moment M. The moment M and the stress S were found to provide the best correlation.

The Arctic sound channel may be an excellent tool for the observation of earthquakes owing to its low noise levels at frequencies less than 15 Hz and its excellent sound transmission characteristic. The Mid Arctic Ridge (83° N, 5° W to 81° N, 120° E) is an active rift zone where earthquakes frequently occur. Keenan and Dyer (1984) observed the stick–slip radiation from these events. The durations can be as long as 100 s and must result from a sequence of emissions. The earthquakes radiate earthborn primary (compressional), secondary (shear) and tertiary waves. The energetic tertiary waves (400 kJ) enter the Arctic sound channels at steep angles and are converted to lower angles by the scattering from the meter-thick ice canopy. The fraction of acoustic energy is large, and durations of 72 s and tertiary wave waterborne arrivals have been observed 300 km from the source in the 5–15-Hz region [see Keenan and Dyer (1984, Fig. 7, p. 823)].

The mathematical description of stick–slip phenomena in the Arctic would seem to be a challenge as one could study earthquakes, ice quakes, microcracking, and a variety of other transient noise sources.

References and Suggested Readings

Introduction

Berlincourt, B. (1981). "Piezoelectric ceramics: Characteristics and applications." J. Acoust. Soc. Am. 70(6): 1586–1595.

Goodmann, R. (2004). A brief history of underwater acoustics in ASA at 75. Acoustical Society of America, Melville, New York, NY: pp. 204–227.

Knudsen, V. O., R. S. Alford, et al. (1948). "Underwater ambient noise." J. Marine Res. 7: 410–429.

Mason, W. P. (1981). "Piezoelectricity, its history and its applications." J. Acoust. Soc. Am. 70(6): 1561–1566.

Spitzer, L. (1945). *Physics of Sound in the Sea*. Peninsula Publishing, Los Altos, CA (1981).

Urick, R. J. (1984). *Ambient Noise in the Sea*. Undersea Warfare Technology Office, NAVSEA, D.O.N., Washington, DC (Also available from Peninsula Publishing, Los Altos, CA, 1986).

Wenz, G. (1962). "Acoustic ambient noise in the ocean: Spectra and sources." J. Acoust. Soc. Am. 34(12): 1936–1956.

Noise Measurement

Bendat, J. S. and A. G. Piersol (1966). *Measurement and Analysis of Random Data*. John Wiley & Sons, New York, NY.

Dyer, I. (1970). "Statistics of sound propagation in the ocean." J. Acoust. Soc. Am. 48(1, pt 2): 337–345.

Horton, C. W. (1968). *Signal Processing of Underwater Acoustic Waves*. U.S. Government Printing Office, Washington, DC (LOCC No.: 74-603409).

Middleton, D. (1987). *An Introduction to Statistical Communication Theory*. Peninsula Publishing, Los Altos, CA.

Pierce, A. D. (1981). *Acoustics: An Introduction to its Physical Principles and Applications*. Acoustical Society of America, Woodbury, NY.

Whalen, A. D. (1971). *Detection of Signals in Noise*. Academic Press, New York, NY.

Infrasonic Noise

Brekhovskikh, L. M. (1967). "Generation of sound waves in a liquid by surface waves." Sov. Phys. Acoust. 12(3): 323–350.
Cato, D. H. (1991a). "Sound generation in the vicinity of the sea surface: Source mechanisms and coupling to the received sound field." J. Acoust. Soc. Am. 89(3): 1076–1095.
Cato, D. H. (1991b). "Theoretical and measured underwater noise from surface orbital wave motion." J. Acoust. Soc. Am. 89(3): 1096–1112.
Goncharov, V. V. (1970). "Sound generation in the ocean by the interaction of surface waves and turbulence." Izv., Atmos. Ocean. Phys. 6(11): 1189–1196.
Hughes, B. (1976). "Estimates of underwater sound (and infrasound) produced by nonlinearly interacting ocean waves." J. Acoust. Soc. Am. 60(5): 1032–1039.
Isakovich, M. A. and B. F. Kur'yanov (1970). "Theory of low frequency noise in the ocean." Sov. Phys. Acoust. 16(1): 49–58.
Kibblewhite, A. C. and K. C. Ewans (1985). "Wave-wave interactions, microseisms, and infrasonic ambient noise in the ocean." J. Acoust. Soc. Am. 78(3): 981–994.
Kibblewhite, A. C. and C. Y. Wu (1989a). "The generation of infrasonic ambient noise in the ocean by nonlinear interaction of ocean surfaces waves." J . Acoust. Soc. Am. 85: 1935.
Kibblewhite, A. C. and C. Y. Wu (1989b). "A reexamination of the role of wave-wave interaction in ocean noise generation." J . Acoust. Soc. Am. 85: 1946.
Latham, G. V. and A. A. Nowroozi (1968). "Waves, weather, and ocean bottom microseisms." J. Geo. Res. 73(12): 3945–3956.
Lloyd, S. P. (1981). "Underwater sound from surface waves according to the Lighthill-Ribner Theory." J. Acoust. Soc. Am. 69(2): 425–435.
Longuet, M. S. (1950). "A theory of the origins of microseisms." Trans. Roy. Soc. A 243: 1–35.
Nichols, R. H. (1981). "Infrasonic ambient noise measurements: Eleuthera." J. Acoust. Soc. Am. 69(4): 974–981.
Urick, R. J. (1984). *Ambient Noise in the Sea.* U.S.G.P.O., Washington, DC (Peninsula Publishing, Los Altos, CA (1986)), pp. 456–963.
Wilson, J. H. (1979). "Very low frequency (VLF) wind-generated noise produced by turbulent pressure fluctuations in the atmosphere near the ocean surface." J. Acoust. Soc. Am. 66(5): 1499–1507 [Cato, D. H. (1981) "Comments on......." J. Acoust. Soc. Am. 70(6): 1783–1784 and Wilson, J. H. (1981) "Rebuttal to......" J. Acoust. Soc. Am. 70(6): 1785–1786.].

Phenomenological Observations of Ambient Noise

Anderson, D. et al. (1972). VLAM Data Analysis-Site 3, Bermuda, Naval Ocean Systems Center, San Diego, CA.
Anderson, V. C. (1979). "Variation of the vertical directionality of noise with depth in the North Pacific." J. Acoust. Soc. Am. 66: 1446–1452.
Arase, E. M. and T. Arase (1965). "Correlation of ambient sea noise." J. Acoust. Soc. Am. 40(1): 205–210.
Axelrod, E., B. Schoomer, et al. (1965). "Vertical directionality of ambient noise in the deep ocean of a site near Bermuda." J. Acoust. Soc. Am. 36(11): 77–83.
Bannister, R. W. (1986). "Deep sound channel noise from high-latitude winds." J. Acoust. Soc. Am. 79(1): 41–48.
Bannister, R. W., R. N. Denham, et al. (1981). Measurement of the Low-Frequency Wind-Generated Ambient Noise in the Deep Ocean, Naval Underwater Systems Center, TD 6565 (Available DTIC A65005).
Buckingham, M. and J. Potter (1995). *Sea Surface Sound'94.* World Scientific, Singapore, 1995.

Burgess, A. S. and Kewley, D. J. (1983). "Wind-generated surface noise source levels in deep water east of Australia." J. Acoust. Soc. Am. 73(1): 201–210.

Carey, W. M. and D. G. Browning (1988). Low-frequency ocean ambient noise, measurement and theory. *Sea Surface Sound*. B. R. Kerman (Ed.), Kluwer Press, Boston, MA, pp. 361–376.

Carey, W. M., R. B. Evans, et al. (1990). "Deep-ocean noise directionality." IEEE J. Ocean. Eng. 15: 324–334.

Carey, W. M. and E. C. Monahan (1990). "Guest Editorial "Special Issue" on sea surface-generated ambient noise: 20-2000 Hz." IEEE J. Ocean. Eng. 15(4): 265–266.

Carey, W. M. and R. A. Wagstaff (1986). "Low-frequency noise fields." J. Acoust. Soc. Am. 80(5): 1523–1526.

Cato, D. H. (1976). "Ambient sea noise in waters near Australia." J. Acoust. Soc. Am. 60(2): 320–328.

Chapman, N. R. and J. W. Cornish (1993). "Wind dependence of deep ocean ambient noise at low frequencies." J. Acoust. Soc. Am. 98: 782–789.

Cox, H. (1973). "Spatial correlation in arbitrary noise fields with application to ambient sea noise." J. Acoust. Soc. Am. 54(5): 1289–1301.

Cron, B. F., B. C. Hassell, et al. (1965). "Comparison of theoretical and experimental values of spatial correlation." J. Acoust. Soc. Am. 37(3): 523–529.

Cron, B. F. and C. H. Sherman (1962). "Spatial-correlation functions for various noise models." J. Acoust. Soc. Am. 34(11): 1732–1736.

Edelblute, D. J., J. M. Fisk, and G. L. Kinnison (1966). "Criteria for optimum signal-detection theory for arrays." J. Acoust. Soc. Am. 41: 199–205.

Fox, G. E. (1964). "Ambient-noise directivity measurements." J. Acoust. Soc. Am. 36: 1537–1540.

Franz, G. J. (1959). "Splashes as sources of sounds in liquids." J. Acoust. Soc. Am. 31(8): 1080–1096.

Furduev, A. V. (1966). "Undersurface cavitation as a source of noise in the ocean." Atmos. Ocean. Phys. 2(235): 314–320.

Garabed, E. P. and R. A. Finkelman (2005). "Measured vertical noise directionality at five sites in the western North Atlantic." IEEE J. Ocean. Eng. 30(2) (Special Issue, Archival Papers): 282–285.

Hollett, R. D. (1992). Observation of Underwater Sound at Frequencies Below 1500 Hz from Breaking Waves at Sea SACLANTCEN Rpt. 183, SACLANT Undersea Research Centre, 19138 San Bartolomeo, Italy.

Hollinberger, D. E. and D. W. Bruder (1990). "Ambient noise data Logger Buoy." IEEE J. Ocean. Eng. 15(4): 286–291.

Kennedy, R. and T. Goodnow (1990). "Measuring the vertical directional spectra caused by sea surface sound." IEEE J. Ocean. Eng. 15(4).

Kerman, B. R. (1984). "Underwater sound generation by breaking waves." J. Acoust. Soc. Am. 75(1): 149–165.

Kerman, B. R. (1988). *Sea Surface Sound*. Kluwer Academic Publishers, Boston, MA.

Kerman, B. R. (1993). *Natural Physical Sources of Underwater Sound*. Kluwer, Dordrecht, The Netherlands.

Kibblewhite, A. C. (1984). "Attenuation of low frequency sound in the sea, Vol. II." J . Acoust. Soc. Am. 75: 645.

Kibblewhite, A. C. and K. C. Ewans (1985). "Wave-wave interactions, microseisms, and infrasonic ambient noise in the ocean." J. Acoust. Soc. Am. 78(3): 981–994.

Kibblewhite, A. C., J. A. Shooter, et al. (1976). "Examination of attenuation at very low frequencies using deep water ambient noise field." J. Acoust. Soc. Am. 60: 1040.

Kibblewhite, A. C. and C. Y. Wu (1989a). "The generation of infrasonic ambient noise in the ocean by nonlinear interaction of ocean surfaces waves." J . Acoust. Soc. Am. 85: 1935.

Kibblewhite, A. C. and C. Y. Wu (1989b). "A reexamination of the role of wave–wave interaction in ocean noise generation." J . Acoust. Soc. Am. 85: 1946.

Knudsen, V. O., R. S. Alford, et al. (1948). "Underwater ambient noise." J. Marine Res. 7: 410–429.

Nystuen, J. A. (1986). "Rainfall measurements using underwater ambient noise." J. Acoust. Soc. Am. 79(4): 972–982.

Perrone, A. (1969). "Deep-ocean ambient noise spectra in the Northeast Atlantic." J. Acoust. Soc. Am. 46(3): 762–770.

Perrone, A. (1976). Summary of a One Year Ambient Noise Measurement Program Off Bermuda, Naval Underwater Systems Center, New London Laboratory, NUSC T.R. 4979 (Available DTIC).

Piggott, C. L. (1964). "Ambient sea noise at low frequencies in the shallow water of the Scotian Shelf." J. Acoust. Soc. Am. 36(1): 2152–2163.

Shang, E. C. and V. C. Anderson (1986). "Surface-generated noise under low wind speed at Kilohertz frequencies." J. Acoust. Soc. Am. 79(4): 964–971.

Shooter, J., T. DeMary, et al. (1990). "Depth dependence noise resulting from ship traffic and wind." IEEE J. Ocean. Eng. 15(4): 292–298.

Shooter, J. and M. Gentry (1981). "Wind generated noise in the Pareca Vela Basin." J. Acoust. Soc. Am. 70(6): 1757–1761.

Talham, R. J. (1964). "Ambient-sea-noise model." J. Acoust. Soc. Am. 36: 1541–1544.

Urick, R. J. (1984). *Ambient Noise in the Sea*. U.S.G.P.O., Washington, DC (Peninsula Publishing, Los Altos, CA (1986)), pp. 456–963.

Wagstaff, R. A. (1978). "Interactive techniques for ambient noise horizontal directionality estimation from towed line-array data." J. Acoust. Soc. Am. 63: 863–869.

Wagstaff, R. A. (1981). "Low-frequency ambient noise in the deep sound channel – the missing component." J. Acoust. Soc. Am. 69: 1009–1014.

Wagstaff, R. A. (2005). "An ambient noise model for the Northeast Pacific Basin." IEEE J. Ocean. Eng. 30(2) (Special Issue Archival Papers): 245–247.

Wales, S. C. and O. I. Diachok (1981). "Ambient noise vertical directionality in the Northwest Atlantic." J. Acoust. Soc. Am. 70: 577–582.

Wenz, G. (1962). "Acoustic ambient noise in the ocean: Spectra and sources." J. Acoust. Soc. Am. 34(12): 1936–1956.

Wenz, G. M. (1972). "Review of underwater acoustics research: Noise." J. Acoust. Soc. Am. 51(2 pt 3): 1010–1024.

Wille, P. C. and D. Geyer (1984). "Measurements on the origin of wind dependent noise variability in shallow water." J . Acoust. Soc. Am. 71(4): 173–195.

Wittenborn, A. F. (1976). Ambient Noise and Associated Propagation Factors as a Function of Depth-Wind-Speed in the Deep Ocean, Tracor Rept. T76RV5060, DTIC (AD 006902), Alexandria, VA.

Worley, R. D. and R. A. Walker (1982). "Low-frequency ambient noise ans sound transmission over a thinly sedimented rock bottom." J. Acoust. Soc. Am. 71(4): 863–870.

Rain Noise

Bom, N. (1969). "Effect of rain on underwater noise level." J. Acoust. Soc. Am. 45(1): 150–156.

Fitzpatrick, H. M. and M. Strasberg (1956). Hydrodynamic Sources of Sound. In *Naval Hydrodynamics*. F. S. Sherman (Ed.), National Academy of Sciences-National Research Council Pub 515, LC-57-60045, pp. 241–280.

Franz, G. J. (1959). "Splashes as sources of sounds in liquids." J. Acoust. Soc. Am. 31(8): 1080–1096.

Guo, Y. P. and J. E. FFowcs-Williams (1991). "A theoretical study of drop impact sound and rain noise." J. Fluid Mech. 227: 345–355.

Heindsman, T. E., R. H. Smith, et al. (1955). "Effect of rain upon underwater noise levels." J. Acoust. Soc. Am. 27: 378–379.

Minnaert, M. (1933). "Musical air-bubbles and sounds of running water." Phil. Mag. XVI (7th Series): 235–249.

Nystuen, J. A. (1986). "Rainfall measurements using underwater ambient noise." J. Acoust. Soc. Am. 79(4): 972–982.

Nystuen, J. A. (1993). An explanation of the sound generated in light in the presence of wind. In *Natural Physical Sources of Underwater Sound.* B. R. Kerman (Ed.), Kluwer Academic Publishers, Boston, pp. 659–668.

Pumphrey, H. C. (1989). "Sources of Ambient Noise in the Ocean: An Experimental Investigation." University of Mississippi, University, MS [Dissertation (NCPA LC.01-1989, Available DTIC)].

Pumphrey, H. C. and L. A. Crum (1989). "Underwater sound produced by individual drop impacts and rainfall." J. Acoust. Soc. Am. 85: 1518–1526.

Pumphrey, H. C. and L. A. Crum (1990). "Free oscillations of near surface bubbles as sources of underwater rain noise." J. Acoust. Soc. Am. 87: 142.

Pumphrey, H. C. and P. A. Elmore (1990). "The entrainment of bubbles by drop impacts." J. Fluid Mech. 220: 539–567.

Arctic Noise

Diachok, O. I. and R. S. Winokur (1974). "Spatial variability of underwater ambient noise at the Arctic ice-water boundary." J. Acoust. Soc. Am. 56(4): 750–753.

Dyer, I. (1983). Song of sea ice and other Arctic ocean melodies. In *Arctic Technology and Policy.* I. Dyer and C. Chryssostomidis (Eds.), Hemisphere Publishing Corporation, New York, NY, pp. 11–37, ISBN 0-89116-361-1.

Dyer, I. and C. Chryssostomidis (1983). *Arctic Technology and Policy.* Hemisphere Publishing Corporation, New York, NY, ISBN 0-89116-361-1.

Ganton, J. H. and A. R. Milne (1965). "Temperature- and wind-dependent ambient noise under midwinter pack ice." J. Acoust. Soc. Am. 36: 406–411.

Greene, C. R. and B. B. Buck (1964). "Arctic ocean ambient noise." J. Acoust. Soc. Am. 36(6): 1218–1220.

Hunkins, K. (1962). "Waves on the Arctic ocean." J. Geophys. Res. 67: 2477–2489.

Keenan, R. E. and I. Dyer (1984). "Noise from Arctic earthquakes." J. Acoust. Soc. Am. 75(3): 819–825.

Macpherson, J. D. (1962). "Some under-ice acoustic ambient noise measurements." J. Acoust. Soc. Am. 34(8): 1149–1150.

Makris, N. C. and I. Dyer (1986). "Environmental correlates of pack ice noise." J. Acoust. Soc. Am. 79(5): 1434–1440.

Mellen, R. H. and H. W. Marsh (1965). Underwater sound in the Arctic ocean. In *Scientific and Engineering Studies: Underwater Acoustics in the Arctic, 1958–1984.* W. A. Von Winkle (Ed.), Naval Underwater Systems Center (Now the Naval Undersea Warfare Center) Newport, RI.

Milne, A. R. (1965). "Statistical description of noise under shore-fast sea ice in winter." J. Acoust. Soc. Am. 39(6): 1174–1182.

Milne, A. R. (1967). Sound propagation and ambient noise under sea ice. In *Underwater Acoustics, Vol. 2.* V. M. Albers (Ed.), Plenum Press, New York, NY, pp. 103–138.

Milne, A. R. and J. H. Ganton (1964). "Ambient noise under Arctic sea ice." J. Acoust. Soc. Am. 36(5): 855–863.

Shepard, G. W. (1979). Arctic ocean ambient noise. *Ocean Engineering.* Massachusetts Institute of Technology, Cambridge, MA.

Urick, R. J. (1984). *Ambient Noise in the Sea.* U.S.G.P.O., Washington, DC (Peninsula Publishing, Los Altos, CA(1986)), pp. 456–963.

VonWinkle, W. A. (1963). "Vertical directionality of deep-ocean noise." J . Acoust. Soc. Am. 35: 1884.

Von Winkle, W. A. (Ed.) (1984). *Scientific and Engineering Studies: Underwater Acoustics in the Arctic, 1958–1984.* Naval Underwater Systems Center (Now the Naval Undersea Warfare Center) Newport, RI.

Yang, T. C., G. R. Giellis, et al. (1987). "Acoustic properties of ice edge noise in the Greenland sea." J. Acoust. Soc. Am. 82(3): 1034–1038.

Chapter 5
Numerical Modeling of Ambient Noise

Introduction

Underwater sound is a first-order compressible pressure oscillation, compared with the zero-order hydrodynamic motions, and propagation is governed by the wave equation. The total sound field is, by the principle of superposition, a linear combination of pressure fields due to individual sources. One or more of the sources may constitute a signal of interest, whereas the rest make up the unwanted noise. In modeling the ambient noise, one tries to calculate the unwanted part of the sound field and its directional characteristics.

The distinction between signal and noise depends on the application. However, one must first decide the characteristics of the sources producing noise. Consider a spatial distribution of monopole sources at one quarter of a wavelength ($\lambda/4$) below the free surface in a range-dependent environment (Carey et al. 1990). The source distribution is a more specific version of the sheet of monopoles used by Kuperman and Ingenito (1980) in a horizontally stratified environment. The near-surface sources will be used to represent shipping as well as wind-driven noise, with a source intensity level per unit area that varies with geographical location. Shipping and wind produce a persistent background noise at a particular place in the ocean. The noise sources do not represent individual ships or breaking wave events, but are represented by an expectation value per unit area. The selection of noise sources also excludes transient sources and is restricted to harmonic sources at low frequency.

Given a set of noise sources, it is necessary to select a model to compute the radiated field from these noise sources. Lateral variability or range dependence is a ubiquitous feature of the underwater sound channel. The preeminent range-dependent underwater acoustic model is the parabolic approximation to the acoustic wave equation due to Tappert (1977) and further developed by Collins (1992, 1993). The parabolic approximation provides a range-marching solution that is essential to the practical calculation of underwater acoustic fields due to distant sources. The assumption of uncoupled azimuths is also essential to make the solution workable on a regional scale. Noise fields are computed, independently, along radials using the two-dimensional parabolic approximation.

W.M. Carey, R.B. Evans, *Ocean Ambient Noise*, Underwater Acoustics,
DOI 10.1007/978-1-4419-7832-5_5,

Once the propagation model has been selected, a choice of ocean acoustic environments needs to be made. This will be done with a particular question in mind: How does sound from near-surface noise sources propagate in the sound channel? The mechanisms responsible for the low-frequency ambient noise have been discussed in a previous chapter and were reviewed by Dashen and Munk (1984), Carey and Wagstaff (1986), and Carey et al. (1990), and are due to radiated noise from ships as well as wind-induced mechanisms (see Chapter 3). These sources, whether over slopes, at high latitudes, or subject to volume scattering, have higher-angle radiation converted to lower angles that propagate within the sound channel. Consider as an example of the first (dominant) mechanism sources over slopes. The calculation depends on the choice of the ocean acoustic environments: bathymetry, sound speed profiles, and bottom types for the particular basin.

The final consideration in a noise calculation is the spatial distribution of the sources or the expected source level per unit area. For radiated noise from ships, the Historical Temporal Shipping (HITS) database (Emery et al. 2001), which contains ship densities by size, is combined with the size-dependent source-level formulas of Renner (1986) to provide the basis of the calculation. Wind source levels are found from satellite wind speed data and the source level versus wind speed curves of Kewley et al. (1990) (also see Appendix I).

Practical considerations, necessary to facilitate the noise calculation for multi-dimensional arrays, are the use of reciprocity to limit the number of sources and spatial interpolation of acoustic fields. Finally, plane wave beam forming is utilized as a method displaying the structure of the ambient noise on a discrete array.

Ambient Noise Modeling

Initial computational models of ambient noise employed ray theory because of its efficiency, availability, and ease of interpretation. The theoretical analysis of Cron and Sherman (1962) and Talham (1964) led to expressions for spatial correlation and vertical directionality, respectively, of surface-generated ambient noise in range-independent environments. Ray theory is a high-frequency asymptotic approximation that degrades at lower frequencies. The wave-theoretic normal-mode approach of Kuperman and Ingenito (1980) extended the results of Cron and Sherman (1962), regarding spatial correlation, to low frequency by the introduction of a subsurface sheet of monopoles.

At higher frequencies (above 1 kHz), the fundamental analyses of Talham (1964) motivated a series of noise models such as the Fast Ambient Noise Model (Cavanagh 1974) and the Ambient Noise Directionality Estimation System [ANDES; Osborne (1979)]. Geometric acoustics, rays, was used in the characterization of the vertical directivity of noise by Cavanagh and Renner (1980), Dashen and Munk (1984), and Harrison (1997a). In addition, geometrical acoustics has led to computational ambient noise codes such as those of Bannister et al. (1989) and Harrison (1997b).

The computational speed of the earlier ray-theoretic codes was retained and combined with aspects of the wave-theoretic approach in the hybrid ray-mode ANDES (Renner 1986) that allowed range-dependent variations in the computational environment. Efficiency was retained in these range-dependent calculations by use of a hybrid ray-mode propagation code and the adiabatic approximation, i.e., the automated signal excess prediction system transmission loss model [ASTRAL; Spofford (1979)] that produced range-averaged propagation loss. The normal-mode component was based on ignoring the phase integral and use of the amplitude of the Wentzel–Kramers–Brillouin (WKB) approximation (Brekhovskikh and Lysanov 2003).

When phase information in the ambient noise field is needed, field calculations become more time-consuming. An attractive method was developed by Hamson and Wagstaff (1983) to handle the need for phase information. Their method is based on adiabatic normal modes and includes modal phase, without recourse to the WKB approximation. A recently published application by Wagstaff (2005) to the Northeast Pacific Basin provides a significant example of the validity of the approach and the ability to compute the directional field characteristics of a large basin. The use of adiabatic normal modes has also been employed by Perkins et al. (1993) for wide ocean area computations.

The advent of the parabolic approximation (Tappert 1977) provided a useful alternative to adiabatic normal modes. The parabolic approximation, although assuming uncoupled azimuths, incorporates intermode coupling in the vertical, which is missing in the adiabatic approximation. Coupling between modes with different vertical angles of propagation is needed to accurately describe downslope conversion and the associated contribution to the vertical directivity of ambient noise. The importance of this effect was shown, computationally, by Carey et al. (1990) using the parabolic approximation of Tappert (1977). The ambient noise model of Hamson and Wagstaff (1983) has been updated to use the parabolic equation technique of Collins (1993) and is described by Breeding et al. (1996).

A comprehensive review of ambient noise modeling can be found in Hamson (1997). Jensen et al. (1994, Chapter 9) provide a complete description of the work of Kuperman and Ingenito (1980), Perkins et al. (1993), and computational results from Carey et al. (1990). Finally, a short review of ambient noise modeling, along with references to earlier reviews, can be found in Etter (2001).

The methods described in this chapter share much in common with those of both Breeding et al. (1996) and Carey et al. (1990). The main difference with the preceding work is that a concerted effort has been made to carefully define every aspect of the ambient noise modeling process, from the input source levels and depths through to the output of the spatial transformation of an array. A second difference is that the code described by Breeding et al. (1996) is often used to model noise due to a discrete distribution of ships; discrete shipping is not considered here.

The Noise Field

Let $P(\omega, \bar{x}, r_j, z_s, \beta_l)$ be the complex acoustic pressure at a receiver position $\bar{x} = col(x, y, z)$ due to a near-surface source at a range $r_j = r_o + j\Delta r, j = 1, J$, depth z_s, and bearing $\beta_l = l\Delta\beta, l = 1, L$. The time dependence $\exp(-i\omega t)$ is neglected, where $\omega = 2\pi f$ and f is the frequency in hertz. The complex pressure values are normalized to be 1 at 1 m from the near-surface sources.

Suppose that the noise source intensity levels $NSL_{j,l}$ depend on location and are specified in units of decibels referenced to $1\mu Pa^2$/Hz at 1 m per square meter of surface area (the reason for using these units is discussed below). The noise source intensity, per unit area, is given by

$$ns_{j,l}^2 = 10^{NSL_{j,l}/10}$$

and is designated as a square of the number $ns_{j,l}$ for convenience. All the noise sources are scaled and combined with random phase at a receiver position to provide a realization of the noise:

$$P_{noise}(\bar{x}) = \sum_{l=1}^{L} \exp(i\psi_l) \sum_{j=1}^{J} \exp(i\psi_j) ns_{j,l} \sqrt{area_j} P(\bar{x}, r_j, z_s, \beta_l), \qquad (1)$$

where ψ_j and ψ_l are uniformly distributed random numbers on the interval $[0, 2\pi]$. The square root of $ns_{j,l}^2 \times area_j$ is used in Eq. (1) to provide units of pressure. The factor $area_j$ reflects the fact that each noise source represents a cell in the range-bearing grid with a different area. The polar range-bearing grid has cells with areas $area_j = \Delta\beta\ \Delta r\ r_j$, as shown in Fig. 5.1.

The intensity level is defined as 10 times the base 10 logarithm of a ratio of intensities (Carey 1995). Such levels are suitable for use in a sonar equation that is based on intensities. For example, the received level is the source level minus the transmission loss. In practice, pressure is measured and intensity is found using the plane wave (oscillatory) approximation, given by $I = |P|^2/2\rho_o c_o$, where ρ_o is density and c_o is sound speed. The intensity level is then

$$10\log_{10}(I/I_{ref}) = 10\log_{10}(|P|^2/p_{ref}^2),$$

where I_{ref} is a reference intensity. When the reference is the intensity in a 1-μPa plane wave, then the intensity level is cited in units of decibels referenced to $1\mu Pa^2$, rather than in watts per square meter. Referencing the intensity level to a pressure squared is a convenient shorthand that will be used here. Similarly, the term "intensity" is used as a shorthand for the term "pressure amplitude squared."

The product of a realization of the noise and its complex conjugate at two positions $\bar{x}$ and $\bar{x}'$ is found from Eq. (1) to be

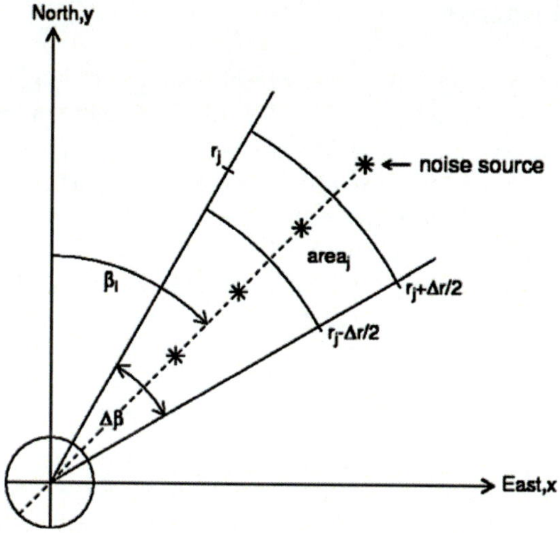

Fig. 5.1 The polar range-bearing grid showing a single sector centered on bearing β_l with width $\Delta\beta$. The sources are indicated by *asterisks*, at the ranges r_j, at the centers of range-bearing cells with area $area_j$. The noise source intensities are proportional to $area_j$. The receivers are clustered in the *circle* at the origin

$$P_{noise}(\bar{x})P^*_{noise}(\bar{x}') = \sum_{l'=1}^{L}\sum_{j'=1}^{J}\sum_{l=1}^{L}\sum_{j=1}^{J}\exp\left[i(\psi_l - \psi_{l'}) + i(\psi_j - \psi_{j'})\right] ns_{j,l}ns_{j',l'}$$
$$\times \sqrt{area_j area_{j'}}P(\bar{x}, r_j, z_s, \beta_l)P^*(\bar{x}', r_{j'}, z_s, \beta_{l'})$$

where the superscript asterisk stands for complex conjugation. The terms with $l = l'$ and $j = j'$ are the same in each realization of the product. Assuming that the noise arriving from different cells ($l \neq l'$ or $j \neq j'$) is uncorrelated, one finds that the average of realizations of products of the noise is approximated as follows:

$$\left\langle P_{noise}(\bar{x})P^*_{noise}(\bar{x}')\right\rangle \cong \sum_{l=1}^{L}\sum_{j=1}^{J} ns^2_{j,l}area_j P(\bar{x}, r_j, z_s, \beta_l)P^*(\bar{x}', r_j, z_s, \beta_l). \qquad (2)$$

The angular brackets represent an average over multiple realizations of the product and is called an "ensemble average." When the noise is stationary, in the statistical sense, the ensemble average can be replaced by a time average. Letting $\bar{x}' = \bar{x}$ in Eq. (2) and taking a logarithm yields

$$10\log_{10}\left\langle |P_{noise}(\vec{x})|^2\right\rangle, [dB\ re : 1\mu Pa^2/Hz] \qquad (3)$$

as the received noise intensity level at $\bar{x}$.

Plane Wave Response

Consider an array consisting of N_p discrete receivers at positions $\bar{x}_n = col(x_n, y_n, z_n)$, $n = 1, N_p$. The plane wave array (or beam) response is based on a time-delayed sum of real pressures, where the time delays are designed to emphasize a plane wave passing across the array from a fixed bearing β and elevation angle α. At a single frequency ω, the Fourier transform of the time-delayed sum is a coherent sum of phase-delayed complex pressure values.

The phase delays are derived from the phase of an incoming plane wave (Pierce 1981):

$$\exp(-i\bar{k} \cdot \bar{x}_n)$$

where $\bar{k} = (\omega/c_o)col(\cos\alpha \sin\beta, \cos\alpha \cos\beta, \sin\alpha)$ is the plane wave vector in Fig. 5.2 and c_o is the speed of sound at the array. The scalar or inner product, in the exponential, is defined as $\bar{k} \cdot \bar{x} \equiv \bar{k}^T\bar{x}$, where the superscript T stands for transpose ($\bar{k}^T$ is a row vector). The sign in the exponential is determined by the time-dependence convention $\exp(-i\omega t)$, mentioned above, and used in the definition of the Fourier transform.

The plane wave noise response can be defined concisely in terms of the column vector

$$\bar{p}_{noise} = col\,[P_{noise}(\bar{x}_n), n = 1, N_p] \tag{4}$$

containing the complex pressure at each receiver and an incoming plane wave replica vector

$$\bar{q}(\alpha, \beta) = col\,[q_n(\alpha, \beta),\ n = 1, N_p], \tag{5}$$

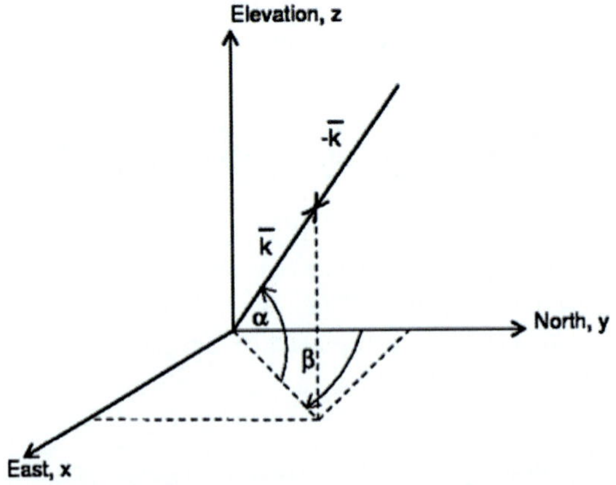

Fig. 5.2 The wave vector $\bar{k}$ plane wave is determined by the elevation angle α and bearing β. The elevation (z) is positive. The incoming wave vector $-\bar{k}$ is also indicated

where $q_n(\alpha,\beta) = \exp(-i\bar{k}\cdot\bar{x}_n)$. Note that the norm squared of the replica vector $\bar{q}(\alpha,\beta)$ is $|\bar{q}(\alpha,\beta)|^2 \equiv \sum_{n=1}^{N_p} |q_n(\alpha,\beta)|^2 = N_p$. The coherent sum of phase-delayed complex pressure values, described above, is the scalar product of the two vectors, in Eqs. (4) and (5), and is given by

$$\bar{q}^*(\alpha,\beta)\cdot\bar{p}_{noise} = \sum_{n=1}^{N_p} q_n^*(\alpha,\beta)P_{noise}(\bar{x}_n). \tag{6}$$

The Cauchy–Schwarz inequality (Stakgold 2000) may be applied in Eq. (6) to obtain

$$\left|\bar{q}^*(\alpha,\beta)\cdot\bar{p}_{noise}\right| \le \left|\bar{q}^*\right| |\bar{p}_{noise}|$$

where equality holds only in the case that the complex pressure vector and plane wave replica vector are scalar multiples of each other. The Cauchy–Schwarz inequality can be restated as an equality:

$$\left|\bar{q}^*(\alpha,\beta)\cdot\bar{p}_{noise}\right| = \cos\theta\ \left|\bar{q}^*\right||\bar{p}_{noise}|$$

where $0 \le \theta \le \pi/2$ is considered to be the angle between the two complex vectors. These observations further justify the use of $|\bar{q}^*(\alpha,\beta)\cdot\bar{p}_{noise}|$ as a measure of the correlation between $\bar{p}_{noise}$ and $\bar{q}^*$. It is desirable to obtain a real scalar with units of pressure amplitude squared, and to this end we define

$$\left|\bar{q}^*(\alpha,\beta)\cdot\bar{p}_{noise}\right|^2 = \left|\sum_{n=1}^{N_p} q_n^*(\alpha,\beta)P_{noise}(\bar{x}_n)\right|^2 \tag{7}$$

as the plane wave response of the array to the noise field.

An equivalent matrix-vector definition of the plane wave response is often used so it is important to present both notations. The matrix-vector definition is obtained from the identity:

$$\left|\bar{q}^*\cdot\bar{p}_{noise}\right|^2 = \bar{q}^H(\bar{p}_{noise}\bar{p}_{noise}^H)\bar{q} \tag{8}$$

where the superscript H stands for complex conjugate transpose. The outer product $\bar{p}_{noise}\bar{p}_{noise}^H$ is the $N_p x N_p$ cross-correlation or covariance matrix of complex pressure on the array. It is also called the cross-spectral density matrix. The average of the plane wave noise response, over realizations, is

$$\left\langle\left|\bar{q}^*\cdot\bar{p}_{noise}\right|^2\right\rangle = \bar{q}^H C_{noise}\,\bar{q} \tag{9}$$

where C_{noise} is the average of the noise cross-correlation matrices, which is defined by $C_{noise} \equiv \langle \bar{p}_{noise}\bar{p}_{noise}^H \rangle$. The trace of the matrix C_{noise} is the sum of the diagonal entries and is given by $trace(C_{noise}) = \langle |\bar{p}_{noise}|^2 \rangle = \sum_{n=1}^{N_p} \langle |P_{noise}(\bar{x}_n)|^2 \rangle$.

It is useful to know an overall bound on the average plane wave noise response in Eq. (7) or equivalently Eq. (9) on the sphere $\bar{q}^H\bar{q} = |\bar{q}|^2 = N_p$. The bound can be obtained by applying the Cauchy–Schwarz inequality to the scalar product on the left-hand side of Eq. (9) and is found to be

$$\bar{q}^H C_{noise}\bar{q} = \left\langle |\bar{q}^* \cdot \bar{p}_{noise}|^2 \right\rangle \le N_p \left\langle |\bar{p}_{noise}|^2 \right\rangle = N_p trace(C_{noise}), \tag{10}$$

where $|\bar{q}|^2 = N_p$.

It is also instructive to know, in some special cases, the actual maximum attained by the average plane wave noise response on the sphere $|\bar{q}|^2 = N_p$. The maximum is obtained from the theory of Hermitian forms [Horn and Johnson (1987), Courant and Hilbert (1953)]. The average noise cross-correlation matrix has the Hermitian symmetry property $C_{noise}^H = C_{noise}$. The Hermitian form, defined in Eq. (9), has nonnegative real values (i.e., it is positive semidefinitive). Consequently, the eigenvalues of C_{noise} are nonnegative real numbers. The sum of the eigenvalues of C_{noise} is also equal to $trace(C_{noise})$. The maximum of the Hermitian form $\bar{q}^H C_{noise}\,\bar{q}$, on the sphere $|\bar{q}|^2 = N_p$, is N_p times the maximum eigenvalue of C_{noise} (when the eigenvalues are characterized in terms of the Rayleigh–Ritz ratio). The maximum of $\bar{q}^H C_{noise}\,\bar{q}$ with $|\bar{q}|^2 = N_p$ will be discussed for two special cases.

In the case of isotropic noise, the noise cross-correlation matrix has the form $C_{noise} = n_{iso}^2 I$, where I is the identity matrix. The maximum eigenvalue (all N_p eigenvalues are equal) is n_{iso}^2. The maximum of the Hermitian form $\bar{q}^H C_{noise}\,\bar{q}$ on the sphere $|\bar{q}|^2 = N_p$ is $N_p n_{iso}^2$. It is noted that the maximum of $\bar{q}^H C_{noise}\,\bar{q}$, divided by N_p, is the average pressure amplitude squared on the array.

In the case that the noise has a persistent dominant component, the average noise cross-correlation matrix is an outer product of a single vector: $C_{noise} = \bar{p}_{noise}\bar{p}_{noise}^H$. The only nonzero eigenvalue of C_{noise} is the scalar $\bar{p}_{noise}^H\bar{p}_{noise} = \sum_{n=1}^{N_p} |P_{noise}(\bar{x}_n)|^2$. The maximum of $\bar{q}^H C_{noise}\,\bar{q}$, on the sphere $|\bar{q}|^2 = N_p$, is $N_p \sum_{n=1}^{N_p} |P_{noise}(\bar{x}_n)|^2$. In this case, the maximum of $\bar{q}^H C_{noise}\,\bar{q}$, divided by N_p squared, is the average pressure amplitude squared on the array. The extra factor of N_p is the difference between the two special cases.

The plane wave array or beam noise intensity level is defined by

$$10\log_{10}\left[\left\langle |\bar{q}^*(\alpha,\beta)\cdot\bar{p}_{noise}|^2 \right\rangle\right] = 10\log_{10}\left[\bar{q}^H(\alpha,\beta)C_{noise}\bar{q}(\alpha,\beta)\right],$$
$$\left[dB\ re: 1\mu Pa^2/Hz\right].$$

In the examples that follow, the plane wave beam noise intensity levels displayed are computed using

$$10\log_{10}\left[\left\langle\left|\bar{q}^{*}(\alpha,\beta)\cdot\bar{p}_{noise}\right|^{2}\right\rangle/N_{p}\right]=10\log_{10}\left[\bar{q}^{H}(\alpha,\beta)C_{noise}\bar{q}(\alpha,\beta)/N_{p}^{2}\right]. \quad (11)$$

The scaling factor $1/N_p^2$ is chosen with the overall bound, in Eq. (10), in mind. The displayed plane wave beam noise intensity levels are bounded, in all cases, by

$$10\log_{10}\left[\left\langle\left|\bar{p}_{noise}\right|^{2}\right\rangle/N_{p}\right]=10\log_{10}\left[trace(C_{noise})/N_{p}\right]; \quad (12)$$

i.e., the displayed noise intensity levels are bounded by $10\log_{10}$ of the spatial average of the omnidirectional noise intensity on the array.

The average omnidirectional noise intensity level in Eq. (12) is the answer to the question: How loud is the noise? The plane wave beam noise intensity level in Eq. (11) provides the structure or directionality of the noise, viewed through an array with a plane wave spatial transform.

Ambient Noise Computation

The direct approach to computing the average cross-correlation matrix C_{noise}, needed in Eq. (11), is to apply the parabolic equation marching procedure to find the complex pressure $P(\omega,\vec{x}_n,r_j,z_s,\beta_l)$ at every receiver for every near-surface source on every bearing. The resulting complex pressure can be used to form the sum in Eq. (2) and to find the entries in the average cross-correlation matrix $C_{noise}^{n,n'}$ defined and approximated by

$$\left\langle P_{noise}(\vec{x}_n)P_{noise}^{*}(\vec{x}_{n'})\right\rangle\cong\sum_{l=1}^{L}\sum_{j=1}^{J}ns_{j,l}^{2}area_{j}P(\vec{x}_n,r_j,z_s,\beta_l)P^{*}(\vec{x}_{n'},r_j,z_s,\beta_l). \quad (13)$$

The direct approach would require $N_p x\, J\, x\, L$ field calculations.
The direct approach is impractical in the range-dependent case, so consider how the number of field calculations may be minimized. Consider a single bearing β_l. The parabolic equation marching procedure is initialized on a subwavelength (10–20 points per vertical wavelength) depth grid: $z_m = m\Delta z$, $m = 1, M$. The starting field is determined by the source depth z_s and the environment at range r_j and bearing β_l. In the direct approach, the field is marched in range, by solving the parabolic equation along a great circle path, through the intervening environments, to each of the receiver positions. The values of the complex pressure $P(\omega,\vec{x}_n,r_j,z_s,\beta_l)$ are then linearly interpolated at the receiver depths, based on the computational depth grid. A single source on a single bearing β_l potentially would require N_p separate parabolic equation calculations. The N_p separate parabolic equation calculations,

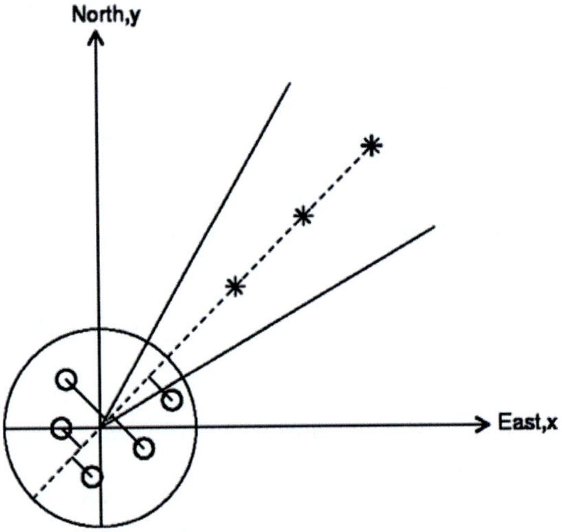

Fig. 5.3 The receiver positions, indicated by *circles*, are projected in the horizontal plane onto the radial through the geometric center of the array

for each source, are avoided by ignoring the cross-range variation of the noise arriving from the sector centered on bearing β_l. Neglecting the cross-range variation is equivalent to projecting the receiver positions, in the horizontal, onto the radial through the geometric center of the array, shown schematically in Fig. 5.3. The projection reduces the number of required calculations to *JxL*.

The projected receiver positions are confined to the vertical plane of a single parabolic equation calculation, shown in Fig. 5.4.

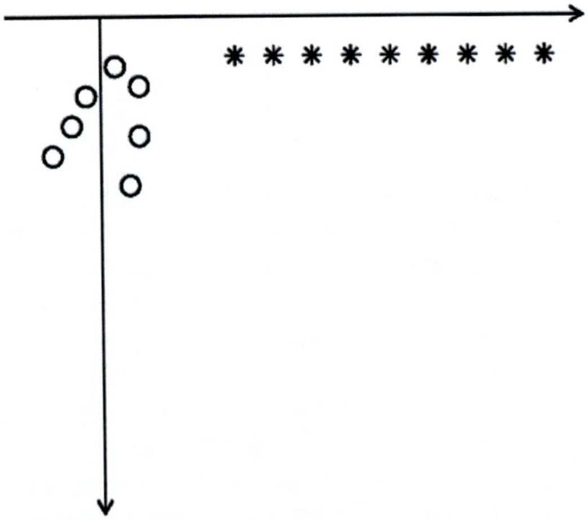

Fig. 5.4 The *circles* indicate the projected receiver positions, in the vertical plane of a single radial. The receivers are below the sea surface and may be offset on either side of the origin, at range zero. The *asterisks* indicate the positions of the near-surface sources

The noise sources are distributed over a large ocean area, so there are usually far fewer receivers than sources ($N_p << J$). Reciprocity may be invoked to further reduce the number of required parabolic equation calculations to $N_p x L$.

Reciprocity allows the sources and receivers to be interchanged, so the receivers are treated as separate sources, and the field, due to each source, is marched out in range, on each radial emanating from the receiving array. Complex pressure values are produced on a depth grid spanning range and bearing for each source in the array. By reciprocity, these complex pressure values are the acoustic field at the receiving array due to sources at every position in the three-dimensional computational grid. Only one or two near-surface receiver depths need be saved for the noise computation. The explicit dependence on the bearing β_l will be suppressed in the following discussion of the field calculations.

Parabolic Equation Field Computations

The reciprocity procedure just described requires that the acoustic field, due to each of the sources, be computed on the same depth and range grid for each bearing. The irregular spacing of the sources, in range, presents an initial obstacle to this requirement that is handled as follows. The parabolic equation is initialized at each of the N_p source ranges. The starting field is a discrete approximation to the exact, delta function, point source solution. It is normalized using the same factor needed to make the free-space point source solution equal to 1 at 1 m from the source. The normalized values of the starting field, on the computational depth grid, due to the nth source at (s_n, z_n) are stored in the column vector

$$\vec{u}(n) = col[u(z_m, n), m = 1, M]. \tag{14}$$

The starting fields are propagated a short distance $r_o - s_n$, along the central radial, to a common range r_o just before the first potential noise source position, shown in Fig. 5.5. This may require a different computational range step for each of the sources. The values of the parabolic equation field, on the computational depth grid, at r_o due to the nth source are similarly stored in the column vector

$$\vec{u}(r_o, n) = P_{mat}(s_n, r_o)\vec{u}(, n), \tag{15}$$

where $P_{mat}(s_n, r_o)$ is a propagator matrix that advances the starting field from range s_n to range r_o. Thereafter, the same computational range step may be used for each of the sources. The parabolic equation field on the depth grid, at range r_j, is stored in the vector

$$\vec{u}(r_j, n) = P_{mat}(r_o, r_j)\vec{u}(r_o, n), \tag{16}$$

where $P_{mat}(r_o, r_j)$ advances the field from range r_o to range r_j.

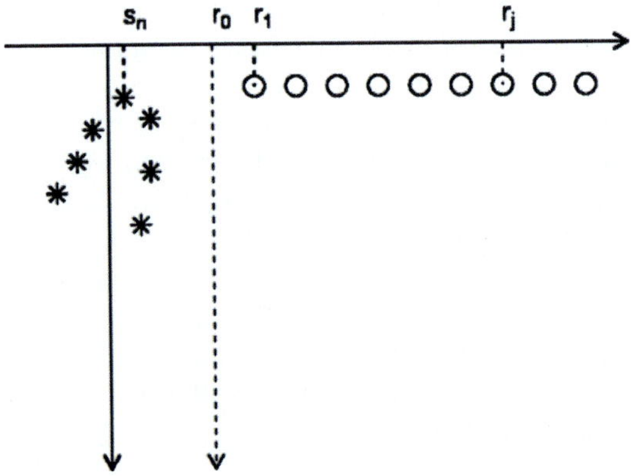

Fig. 5.5 Reciprocity is invoked to interchange the sources and receivers. The ranges of the sources are indicated by s_n. The parabolic equation field, due to each of the sources, is advanced to range r_o, whereupon a single fixed-range step may be used to advance the field to subsequent ranges r_j

The propagator matrix is a discretization of the exponential of a depth-separated differential operator. The propagator matrices are not actually computed since the computation involves finding the inverses of many *MxM* tridiagonal matrices. The effects of the propagator matrices are computed by the solution of a sequence of tridiagonal systems, advancing the *Mx*1 parabolic equation field vector from one range to the next. All of the ranges r_j for one source, on one bearing, are computed in a single sweep, out to the maximum range r_J. The process is repeated, in parallel, for each of the sources $n = 1, N_p$. After range r_o, the parallel implementation becomes more efficient since the tridiagonal coefficient matrices are independent of n, effectively advancing an MxN_p parabolic equation field matrix, with entries $u(r_j, z_m, n)$, from one range to the next.

The complex pressure is obtained by restoring the carrier wave and cylindrical spreading, from the nth source, to the parabolic equation field

$$P(s_n, z_n, r_j, z_s) = \exp[ik_o(r_j - s_n)]u(r_j, z_{ms}, s_n, z_n) \Big/ \sqrt{(r_j - s_n)}, \tag{17}$$

where the index n has been replaced by (s_n, z_n). The point z_{ms} is the depth in the computational grid that is nearest to z_s. The value $k_o = \omega/c_o$ is the reference wavenumber. The complex pressure values $P(\vec{x}_n, r_j, z_s, \beta_l)$, in Eq. (13), are obtained from Eq. (17) by identifying (s_n, z_n) as the projection of $\vec{x}_n = (x_n, y_n, z_n)$ and recognizing that both $s_n = x_n \sin\beta_l + y_n \cos\beta_l$ and $u(r_j, z_{ms}, s_n, z_n)$ depend on the bearing β_l.

The lack of a closed-form representation for the parabolic equation field $u(r_j, z_{ms}, s_n, z_n)$ makes the evaluation of Eq. (13) based on Eq. (17) intuitively unsatisfying and provides very little insight into the form of the entries in the average noise cross-correlation matrix. This situation is partially rectified if one accepts the assumption of horizontal stratification as done by Kuperman and Ingenito (1980).

Their modal approach is applied in Appendix J to provide some insight into the nature of the average noise cross-correlation matrix.

Field Interpolation

When the receivers in the array move a small distance (a few wave lengths) it becomes necessary to repeat the entire noise calculation or apply some kind of interpolation scheme. The commonly used modal interpolation technique (Heany 2009) requires the calculation of the acoustic field on a very fine mesh over the entire computational depth interval, at the geographic center of the array. The acoustic field is projected onto the local normal modes, with assumption that the environment is approximately horizontally stratified under the array. The resultant modal expansion can be used to interpolate the acoustic field on the array, but the technique does not afford the advantages of applying reciprocity.

When reciprocity is used, the source positions are the receiver positions in the array. This is done as a matter of efficiency since there are far fewer receivers than noise sources. It is also possible to save the field on the array, due to each noise source. An interpolation procedure, which retains the same level of efficiency, is based on choosing source positions to span only the receiving array depth and range interval, as shown in Fig. 5.6. Different distributions of receivers can be accommodated by interpolation. The acoustic field is a smooth function of depth within the water column and may be accurately interpolated using Legendre or Chebyshev polynomials with as few as π (three or four) points per vertical wavelength (Gotlieb and Orszag 1977). The source depths $d_{nz}, nz = 1, N_z$ are chosen as the zeroes

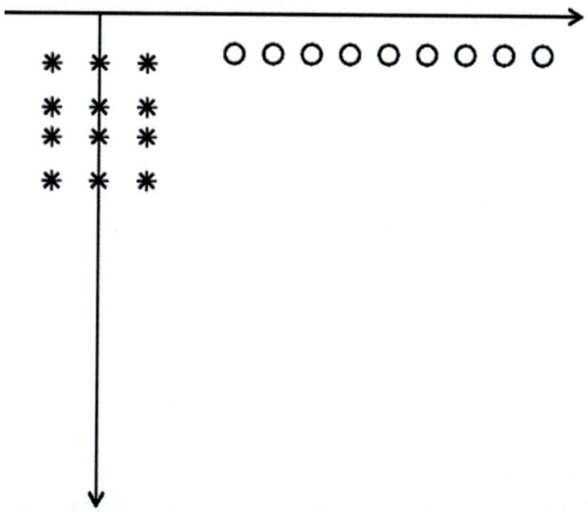

Fig. 5.6 A regular distribution of source positions, based on the zeros for the Legendre polynomials, is used to interpolate the field, due to an irregular set of positions (e.g., Fig. 5.5)

of the N_zth-order Legendre polynomial in the depth interval. The source ranges $s_{nr}, nr = 1, N_r$ may be chosen uniformly, in range, at a rate of one or two points per horizontal wavelength since the parabolic equation field is a slowly varying function of range. The number of sources is $N_z x N_r$.

The parabolic equation fields at the receiver positions $(z_n, s_n), n = 1, N_p$ are found by polynomial interpolation in depth and linear interpolation in range. A weighted sum of the parabolic equation field values at the depths $d_{nz}, nz = 1, N_z$ is used to find the expansion coefficients of the field, in terms of the first $(N_z - 1)$ Legendre polynomials, at each range $s_{nr}, nr = 1, N_r$. The weighted sum is an implementation of Gaussian quadrature (Press et al. 1992). The resulting polynomial coefficients are interpolated linearly in range at $s_n, n = 1, N_p$. The Legendre polynomial expansion, with interpolated coefficients, is evaluated at the receiver depths $z_n, n = 1, N_p$.

Noise Realizations

The field calculations discussed so far are equally well suited for generating replica fields for use in matched field processing (Bucker 1976). If one cannot afford the luxury of storing the fields due to individual sources, it is possible to superimpose the fields due to all the near-surface sources, on a radial, in a single parabolic equation calculation without invoking reciprocity (Carey et al. 1990). The superposition is coherent, so it is necessary to randomly select phases for each source and form realizations of the noise field using the sum in Eq. (1). If N_{rel} realizations of the noise field are required to estimate the average over realizations in $C^{n,n'}_{noise}$, then $N_{rel} x L$ parabolic equation calculations are needed. Experience has shown that 20 or 30 realizations are sufficient to obtain a representative average. A study of the behavior of the variance of the realizations can be found in Dyer (1973).

The superposition procedure is justified by substituting the complex pressure field in Eq. (17) into Eq. (1) to obtain

$$P_{noise}(\vec{x}_n) = \sqrt{\Delta\beta\Delta r} \sum_{l=1}^{L} \exp[i(\psi_l - k_o s_{n,l})] \sum_{j=1}^{J} \exp[i(\psi_j + k_o r_j)] n s_{j,l} u(r_j, z_{ms}, s_{n,l}, z_n, \beta_l), \tag{18}$$

where $\psi_j + k_o r_j$ can be considered to be uniformly distributed on the interval $[0, 2\pi]$, since the ranges of the near-surface sources may be randomly perturbed by as much as a wavelength. The bearing dependence of the projected range $s_{n,l}$ is indicated by the subscript l. The factor $\exp[-ik_o s_{n,l}]$ is persistent over realizations and contains the gross interelement phase structure of the noise arriving from bearing β_l, which is common to the entire array. The rest of the phase information is contained in $u(r_j, z_{ms}, s_{n,l}, z_n, \beta_l)$. In Eq. (1), the factor $\sqrt{area_j}$ cancels the cylindrical spreading factor in Eq. (17), assuming that all the $s_{n,l}$ are small compared with r_j.

The sum over the ranges r_j in Eq. (18) can be found by starting the parabolic equation calculation at the most distant near-surface source (z_s, r_J) on the radial, using the geometry shown in Fig. 5.4. The parabolic equation is marched toward the receiving array, superimposing additional near-surface sources as they arise at (z_s, r_j) (Carey et al. 1990). Each new source is randomly phased and provided with the appropriate amplitude $ns_{l,j}$, depending on its location. The same computational range step may be used on the entire range interval. The procedure is stopped when the accumulated parabolic equation field is marched past the receiving array. The accumulated sum, at $(s_{n,l}, z_n)$, is found using bilinear interpolation in the computational range-depth grid. Multiple realizations of the sum over range could be obtained in a parallel implementation of the parabolic equation marching procedure. This approach is not applied in the following examples.

Environmental Considerations

The parabolic equation calculations, described above, require extraction of environmental data along radials. On a spherical earth, radials are great circle paths. The data that need to be extracted are bathymetry, sound speed profiles, and bottom descriptions, including sound speed, density, and attenuation. The near-surface noise sources invariably interact with the bottom, even in deep water. Consequently, knowing the angular dependence of the bottom reflectivity over large areas is important. When great distances are involved, for ambient noise calculations at low frequencies, volume attenuation of seawater is also important. Otherwise, noise intensity levels continue to increase with range beyond physically realistic levels (also see Appendix J).

Another aspect of long-range sound propagation is the curvature of the underwater acoustic waveguide. The acoustic path along the bottom of the ocean is shorter than the acoustic path along the surface. This earth curvature effect can be included in the parabolic equation calculations by increasing the sound speed toward the bottom of the ocean by multiplying the sound speed by $1 + z/rad_e$ where rad_e is the radius of the spherical earth and z is depth. This is a simplification of the earth curvature correction found in Tappert (1977).

Noise Source Levels

The noise source intensities $ns_{j,l}^2$ need to be defined and the surface noise source depth $z_s = \lambda/4$ will be justified. Once this has been accomplished, the ambient noise modeling objectives of determining the average omnidirectional noise intensity level and the plane wave noise intensity levels or noise directionality will, theoretically, be accomplished. The definitions of $ns_{j,l}^2$ and r_s are broken down into two cases: wind noise and shipping.

Wind Source Levels

The definition of the wind noise source levels is predicated on the assumption that breaking waves produce wind noise. Breaking waves cause a surface impact, which is a dipole, and an oscillating bubble cloud at depth, which is also like a dipole owing to its proximity to the surface. Either may be approximated by a near-surface monopole, as has already been assumed in the modeling description. A comprehensive summary of wind noise source levels, obtained from a wide range of experimental measurements, was reported by Kewley et al. (1990) (also see Appendix I). In some cases, the measured data were directly inverted for monopole source levels at a depth of one quarter of a wavelength, and in other cases the measured source levels were converted to monopole source levels at a depth of one quarter of a wavelength. The motivation for using a quarter of a wavelength is that deeper sources are no longer dipolelike, whereas shallower sources present modeling difficulties. Subsequent measurements by Chapman and Cornish (1993) used the same monopole source depth convention.

The wind noise source level curves reported by Kewley et al. (1990) are summarized in Fig. 5.7 in units of decibels referenced to $1\,\mu Pa^2/Hz$ at 1 m per square meter of surface area. Figure 5.7 shows monopole source levels $WSL\,(f,u)$ for a source at a depth of one quarter of a wavelength as a function of frequency f in hertz and wind speed u in meters per second. Assume that a database of wind speed is available, organized by latitude and longitude. Given the latitude and longitude of the receiving array and a range r_j and bearing β_l from the array, the wind speed $u_{j,l}$ can be extracted from the database. The noise source intensities are

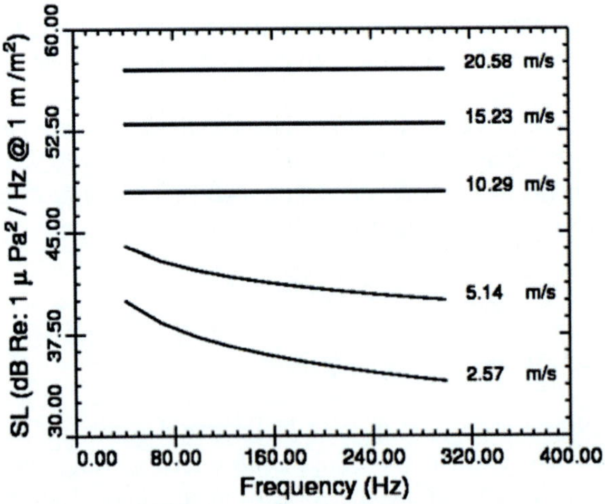

Fig. 5.7 Wind noise source intensity level curves versus frequency and wind speed, due to Kewley et al. (1990). The wind speeds in meters per second are equivalent to 5, 10, 20, 30 and 40 knots, respectively. The source levels are in units of decibels referenced to $1\,\mu Pa^2/Hz$ at 1 m per square meter of surface area. The source depth is assumed to be one quarter of a wavelength

$$ns^2_{j,l} = 10^{WSL(f,u_{j,l})/10}, \tag{19}$$

where $WSL(f, u_{j,l})$ is found using bilinear interpolation in the table of discrete values that are plotted in Fig. 5.7. The frequency is the same as the one used in the parabolic equation calculations. The near-surface source depth, for wind noise, is $z_s = \lambda/4$, where $\lambda = c_o/f$ is the wavelength.

Shipping Source Levels

The primary noise generation mechanism for ships is cavitation at the propeller. The depths of the propellers are between 3 and 12 m (Gray and Greely 1980). The hull and the wake of the ship are important in the horizontal directionality of this radiated noise for discrete ships, but it is assumed to average out when ship densities are used. The average dead weight tonnage of merchant ships increased by one third between 1980 and 1995 (Maritime-Administration 1995), so it is suggested that the near-surface source depth r_s be selected on the basis of tonnage to be between 6 and 12 m. Also, see the article by Ross (2005) regarding a potential increase in shipping source levels. A source depth of r_s= 7.5 m, or one quarter of a wavelength, is used in the 50-Hz shipping noise example that follows.

The method for defining the noise source intensity levels is driven by the data in the shipping database HITS 4.0 (Emery et al. 2001), which contains ship densities by ship type, organized by latitude and longitude. The ship types, $st = 1, 5$, are supertankers, large tankers, merchants, tankers, and fishing vessels, in order of decreasing average source level. The ANDES (Renner 1986) curves provide the source levels $SSL(f, st), st = 1, 5$, in units of decibels referenced to $1\mu Pa^2/Hz$ at 1 m, as a function of frequency for the same five ship types (see Fig. 5.8). The ANDES source levels were based on a monopole point source at a depth of 6 m.

Given the latitude and longitude of the receiving array and a range r_j and bearing β_l from the array, the ship densities $d_{j,l}(st), st = 1, 5$ by ship type, in number of ships per 1,000 square nautical miles, can be extracted from the HITS database. The noise source intensities $ns^2_{j,l}$ are found using

$$ns^2_{j.,l} = (1.852)^{-2}x10^{-9} \sum_{st=1}^{5} d_{j,l}(st)10^{SSL(f,st)/10}, \tag{20}$$

where the scaling factor converts 1/(1,000 nm^2) to 1/m^2, and $SSL(f, st)$ is obtained from Fig. 5.8.

A more sophisticated model of shipping source levels can be obtained on the basis of length and speed using results of Ross (1987) as implemented by Hamson and Wagstaff (1983). The shipping source level curves of Renner (1986) are used here for direct compatibility with the HITS ship types.

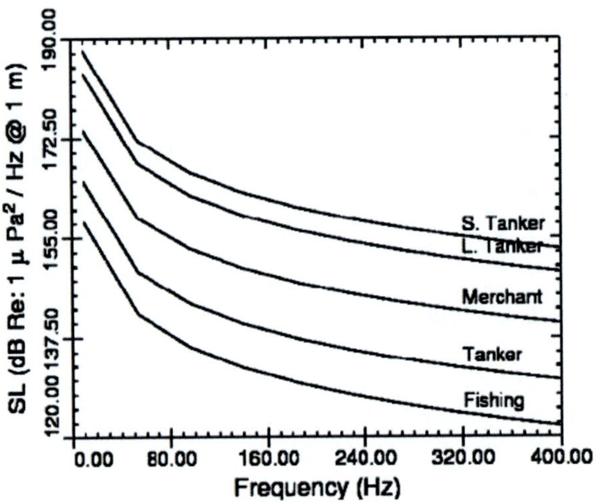

Fig. 5.8 Ambient Noise Directionality Estimation System (ANDES) shipping source intensity level curves versus frequency and ship type, due to Renner (1986). The source levels are in units of decibels referenced to $1\mu Pa^2/Hz$ at 1 m. The source depth was assumed to be 6 m

Ambient Noise Examples

The noise modeling technique that will be used here is the evaluation of Eq. (13), on a sector-by-sector basis, with the assumption that noise arriving from different sources is summed incoherently. The complex pressure values, needed in Eq. (13), are supplied by the parabolic equation fields in Eq. (17), computed on a radial-by-radial basis. This approach will be applied to a set of illustrative examples.

An array site at 54° N, 145° W in the Gulf of Alaska is chosen since it exhibits characteristics of an enclosed basin as well as the open ocean, depending on the bearing. The ranges to the boundaries of the basin to the north are 600–700 km, and calculations as far away as 3,500 km to the south are employed (the maximum range interval will be justified subsequently). In the absence of attenuation, ambient noise levels grow without bound as more distant sources are added. A physical limit is imposed by the volume attenuation of seawater. The appropriate value for the North Pacific is quite small: 4×10^{-6} dB/wavelength (Lovett 1980). Bathymetric contours of the region are shown in Fig. 5.9. The range rings are 500 km apart and are plotted on a Mercator projection. A frequency of 50 Hz is used to minimize the computational burden.

The receiving array geometry affects the plane wave noise response. A combination of horizontal and vertical array apertures substantially complicates the interpretation of the noise directionality. Consequently, a simple vertical array is used. The azimuthal directionality of the noise field is obtained, in the calculation, using Eq. (11) on a sector-by-sector basis, including only noise sources in

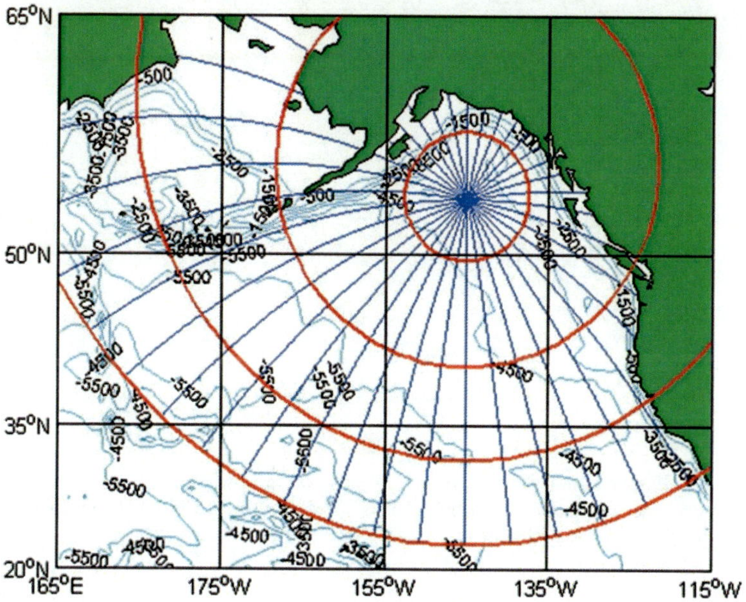

Fig. 5.9 Array site at 54° N, 145° W in the Gulf of Alaska showing elevations in meters to a range of 3,500 km. The radials are great circle paths, plotted on a Mercator projection. Field calculations are stopped when the water depth is a fraction of a wavelength. The range rings start at 500 km and are 1,000 km apart

the individual sectors. The array consists of 41 receivers spaced 7.5 m apart in the depth interval 50–350 m.

The area around the array is characterized by sound speed profiles from the Generalized Digital Environmental Model (GDEM) and bathymetry from the Digital Bathymetric Data Base (DBDB). Both databases were available from the Naval Oceanographic Office (2004). Sound speed profiles, for the fall season, and bathymetry were extracted from GDEM and DBDB at 10-km intervals, along 36 radials at 10° increments out to a range of 3,500 km. The sound channel axis is at a depth of 100 m at the array site, in 4,000 m of water. The sound speed profile at the array is shown in Fig. 5.10.

The availability of bottom profiles and bottom type is limited because of the lack of the availability of standard databases. This issue can be avoided by using a silty-clay bottom (Hamilton 1980) in the basin and a sandy-silt bottom on the slopes at depths shallower than 2,500 m. The depth 2,500 m was taken as the beginning of the continental slope. The critical angles of these bottoms are approximately 0° and 20°, with the slopes being more reflective. The sound speed profiles, bottom profiles, and bathymetry were used to generate input files for the parabolic equation model representing 36 sectors with a width of 10°. The parabolic equation model was run in the reciprocal mode with 10 sources in the 50–350-m depth interval. The

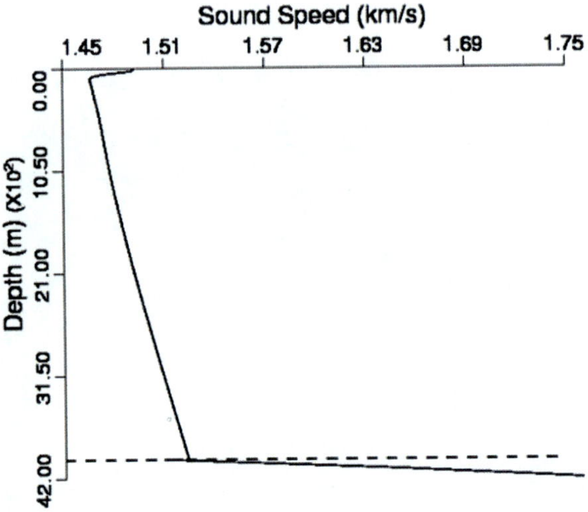

Fig. 5.10 Sound speed profile at 54° N, 145° W in 4,000 m of water, along with the first 200 m of the silty-clay seabed. The sound channel axis is at 100 m at this latitude. The receiving array is vertical, extending from 50 to 350 m

sources were located at the zeroes of the 41st-degree Legendre polynomial to allow interpolation in depth.

Uniform Noise

The thing that distinguishes the various examples is the near-surface noise source distribution. A baseline case uses a uniform distribution of noise sources in range and bearing. It is designed to emphasize the propagation effects alone. The uniformly distributed sources are given a source level of 44 $dB\ re(1\mu Pa)^2/Hz@1m/m^2$ corresponding to a 10-knot wind speed, in Fig. 5.7. The baseline case is not meant to be a realistic problem, but removes source distribution irregularities and simplifies interpretation. The noise sources are placed at a depth of 7.5 m (about $\lambda/4$ at 50 Hz), at a range increment of 100 m, on the range interval 0.1–3,500 km. There are 35,000 sources on radials extending to the south.

All the components needed to perform a noise calculation for the baseline case with a uniform distribution of noise sources have been specified. The average omnidirectional received level for this case, computed using Eq. (12), is 63.4 $dB\ re : 1\mu Pa^2/Hz$. The average omnidirectional level changes by less than 0.1 dB with the introduction of the sources in the range interval between 2,500 and 3,500 km. The range limit is justified by the overall attenuation in the waveguide and the average sound speed structure in the North Pacific. In Eq. (12), the received level also was computed on a sector-by-sector basis to find average (sectorwise) omnidirectional levels for sources in each of the 36 sectors. The sectorwise omnidirectional levels versus bearing are plotted in Fig. 5.11 along with the average omnidirectional level.

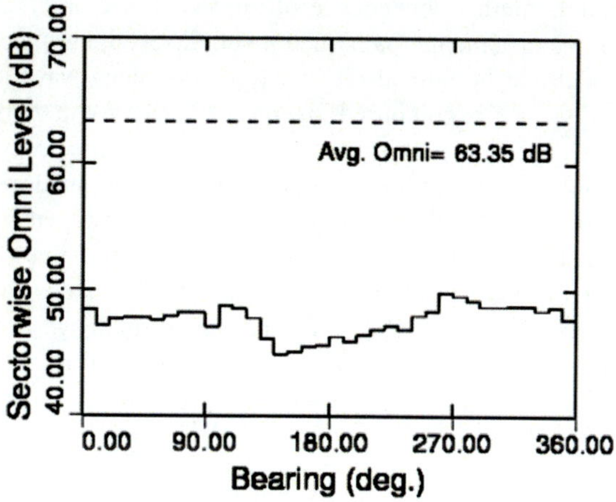

Fig. 5.11 The sectorwise omnidirectional noise levels versus bearing for the uniform distribution of sources, with a source intensity level of 44 $dB\ re(1\mu Pa)^2/Hz@1m\ /m^2$. The average omnidirectional received intensity level is indicated by the *dashed line*

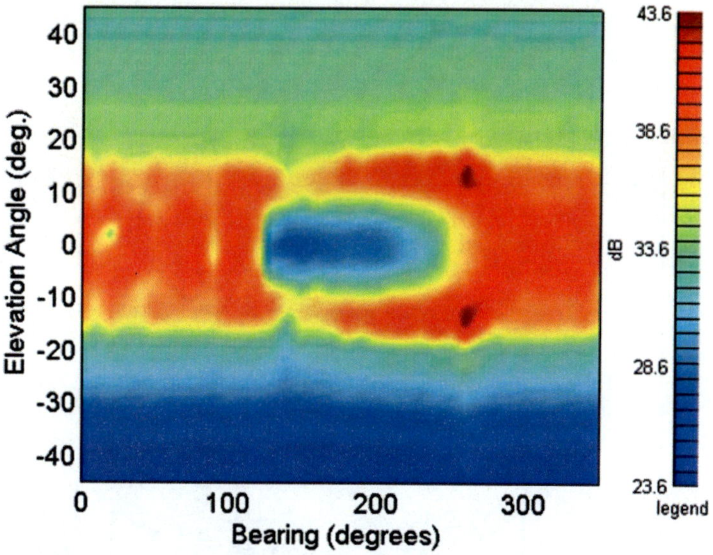

Fig. 5.12 The plane wave noise response versus bearing and elevation angle for the uniform distribution of sources. The azimuthal dependence of the vertical directivity is obtained on a sector-by-sector basis. The downslope conversion of the noise from the sources over the slopes, to the north, fills in the notch at the horizontal. The directivity of the noise arriving from the open ocean, to the south, peaks at ±15°

There is a mild azimuthal dependence of the sectorwise omnidirectional levels owing to different environments (bathymetry, sound speed, etc.) along each of the radials. The vertical and horizontal directivity of the plane wave noise response, computed using Eq. (11), is contoured in Fig. 5.12, where the vertical angles run between +45° and −45°. The peak response in the vertical directivity on each of the radials is less than the corresponding sectorwise omnidirectional levels in Fig. 5.11.

The vertical directivity of the plane wave noise response peaks broadly at the horizontal for radials that intersect the continental slope to the north, and peaks at ±15° for radials to the south. The broad peak is due to downslope conversion of the sources over the slope (Wagstaff 1981). The arrivals from above 20° are due to the nearest near-surface sources starting 100 m away from the array.

Wind Noise

For the uniform source distribution, the general characteristics of the plane wave noise response are disrupted partially by a nonuniform source distribution. An example of a nonuniform source distribution derived from a measured wind speed field and based on the wind noise source level curves in Fig. 5.7 is shown in Fig. 5.13. The uniform level of 44 dB, used in the baseline case, falls into the range of wind noise source levels, but there are higher source levels to the west and southwest as well as lesser wind noise source levels elsewhere. The

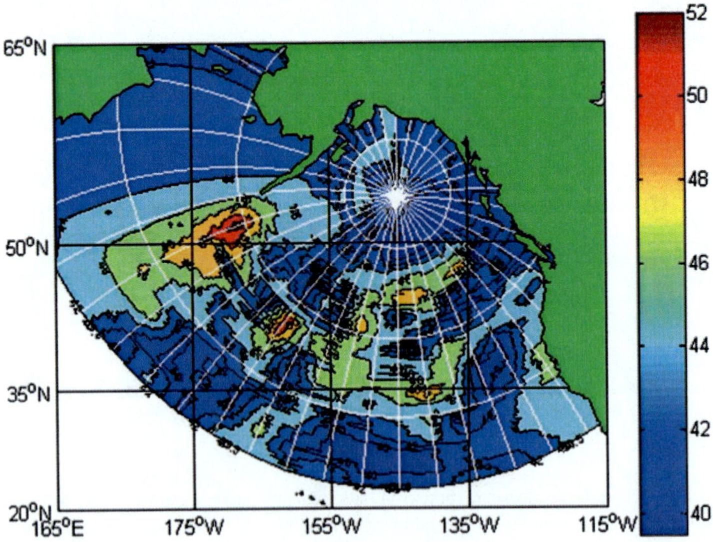

Fig. 5.13 Wind noise source levels in units of decibels referenced to $1\mu Pa^2/Hz$ at 1 m per square meter obtained from satellite measurements combined with the wind noise source level curves in Fig. 5.7. The geographic region is the same as in Fig. 5.9

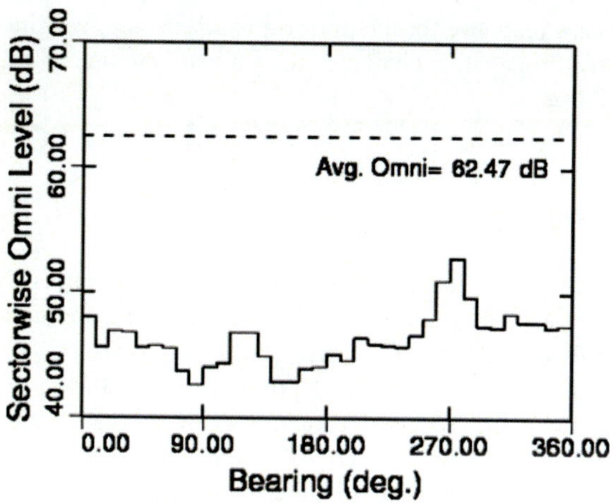

Fig. 5.14 The sectorwise omnidirectional wind noise levels versus bearing for the measured wind speed case. The source levels for this case are shown in Fig. 5.13. There is a peak at 270°, due the higher wind source levels on the radial that grazes the Aleutian Islands chain

average omnidirectional level for the measured wind speed case is within 1 dB of the baseline case.

The horizontal variability of the sectorwise omnidirectional levels, shown in Fig. 5.14, is greater, as expected, with a peak at around 270° in the direction of the higher wind noise source levels in Fig. 5.13. The plane wave noise response, due to the observed winds, in Fig. 5.15, displays the horizontal directionality of

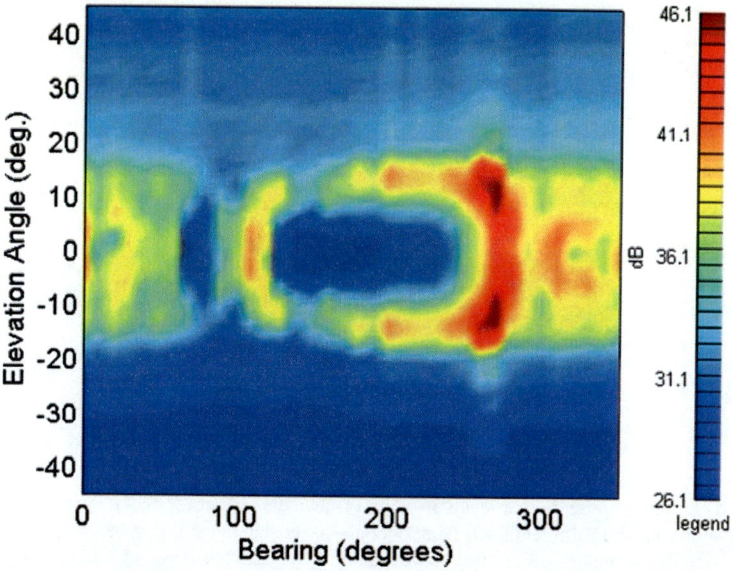

Fig. 5.15 The plane wave noise response versus bearing and elevation angle for the measured distribution of wind noise sources. The peak response is at a bearing of 270°

the plane wave noise response for the uniform baseline case, weighted by the wind noise source levels. Propagation characteristics already provide an enhanced weight to sources over slopes.

Shipping Noise

A second example of a nonuniform source distribution is derived from the HITS database (Emery et al. 2001) and is based on the shipping source level curves in Fig. 5.8. The distribution of shipping source levels is shown in Fig. 5.16. The source levels are applied to a point source at a depth of 7.5 m, just as in the case of wind noise. A quick comparison of Figs. 5.13 and 5.16 shows that shipping is the dominant source of ambient noise at 50 Hz. All other things being equal, there would only be a noise intensity level and spatial distribution difference between wind and shipping ambient noise. This is not the case since ships are fewer and farther apart than breaking wave events.

The shipping noise sources are excluded for an area with a radius of 350 km from the receiving array and are spaced 10 km apart in the interval between 350 and 3,500 km. Nearby ships are not part of the background ambient noise since they may be resolved individually. The sectorwise omnidirectional levels versus bearing are plotted in Fig. 5.17 along with the average omnidirectional level of

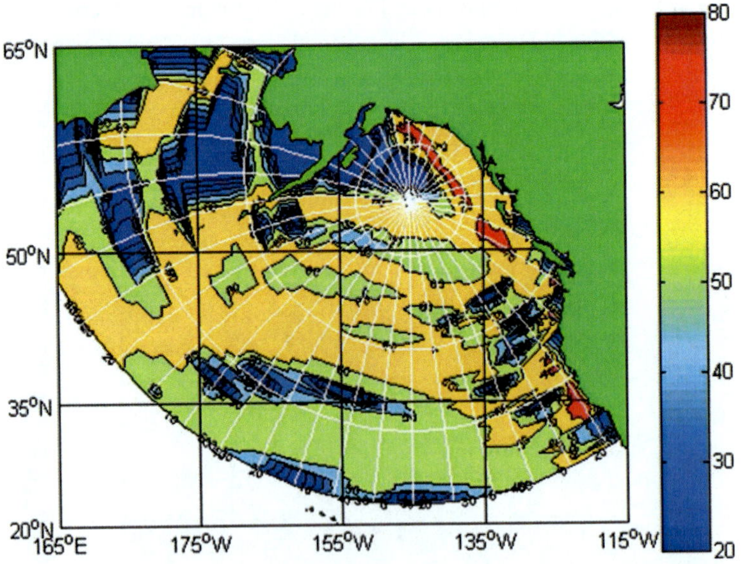

Fig. 5.16 Shipping noise source levels in units of decibels referenced to $1\mu Pa^2/Hz$at 1 m per square meter obtained from HITS 4.0 (Emery et al. 2001) ship densities, by ship type, combined with the ANDES shipping source level curves in Fig. 5.8. Since both Fig. 5.13 and this figure use the same source level units, it is apparent that shipping noise dominates wind noise at 50 Hz

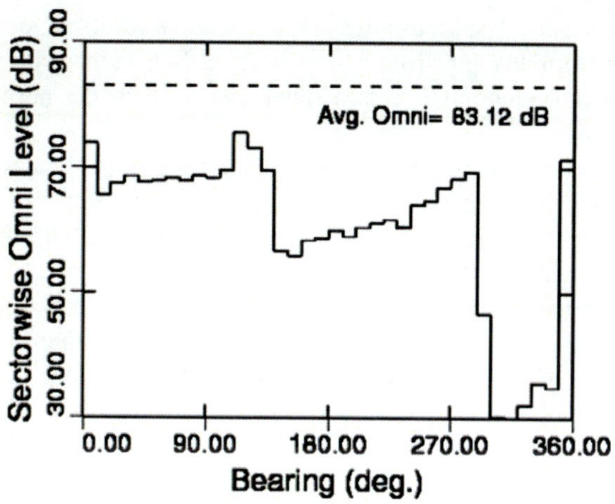

Fig. 5.17 The sectorwise omnidirectional wind noise levels versus bearing for shipping noise based on the source levels in Fig. 5.16. The average omnidirectional shipping noise intensity level is 20 dB higher than the average omnidirectional wind noise intensity level in Fig. 5.14. Ships are excluded from the first 350 km around the array. The peaks at 0° and 120° correspond to shipping lanes that intersect the continental slope near San Francisco, California, and Anchorage, Alaska

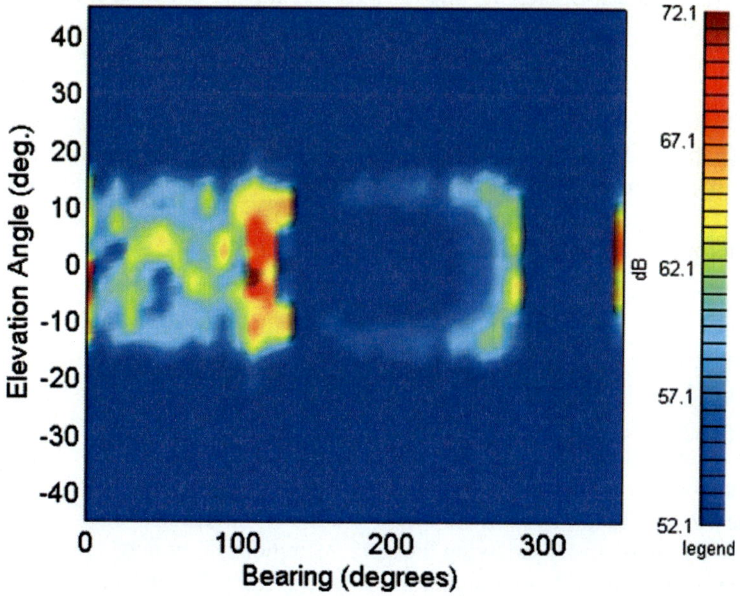

Fig. 5.18 The plane wave noise response versus bearing and elevation angle for distant shipping. The uneven nature of the vertical directivity, to the northeast, may be smoothed by distributing the shipping sources more densely and reducing the source levels proportionally

83.1 $dB\ re: 1\mu Pa^2/Hz$. The sectorwise omnidirectional levels are peaked at 0° and 110°, where the ships leaving from and arriving at San Francisco and Anchorage cross the continental slope. There is a quiet quadrant, to the northwest, due an absence of shipping

The plane wave noise response, due to the distant shipping, is shown in Fig. 5.18. The vertical directivity of the shipping noise, to the northeast, reflects the distribution of shipping over the continental slope. As is the case with the wind noise, the vertical directivity of the shipping noise, to the south, peaks at ±15°. The 10-km-range increment between shipping noise sources does not have a strong effect. Using a 100-m increment between the shipping noise sources smoothes out vertical directivity of the shipping noise in Fig. 5.18, but does not significantly change its interpretation.

Summary of Examples

Each of the three examples demonstrates that energy from near-surface sources, over slopes, gets into the sound channel. The downslope conversion of energy from both wind and shipping, to the north, is obviously greater in the directions where the source levels are greater. The shipping dominates the wind at 50 Hz.

The examples show the classic vertical directivity of the noise, to the south. The near-surface sources excite only steeply traveling waves. The steepest are stripped away by bottom interaction, leaving only a small group of modes arriving at ±15°. Other mechanisms, which have the potential to convert these modes into shallower angles, were not included in the parabolic equation modeling.

Both internal wave scattering (Dozier and Tappert 1978) and rough surface scattering (Beilis and Tappert 1979) can result in an equipartitioning of modal energy, over sufficient ranges. It follows that if internal wave and rough surface effects were included in the parabolic equation calculations, then some filling of the noise notch at the horizontal could be expected.

Summary

In this chapter, the approach has been to carefully define each aspect of the ambient noise modeling process, within the context of the assumptions that have been made. The components in the modeling process, such as the input source levels and depths and the acoustic field calculations, take a modular form. These components may be replaced by alternatives, satisfying the same definitions, to provide improved noise modeling capabilities. For example, the shipping source levels are candidates for improvements, as mentioned already.

The set of definitions introduced in this chapter will yield a normalized prediction of the average noise cross-correlation matrix, and the plane wave beam noise levels are shown in the summary plate. The noise cross-correlation matrix is normalized so that the average of diagonal entries is the average omnidirectional noise intensity on the array. The average omnidirectional noise level is $\mathrm{NL_o} = 10\log_{10}\left[trace(C_{noise})/N_p\right]$, dB re 1 μPa^2/Hz. The plane wave beam noise level characterizes the directionality of the ambient noise.

Summary

Average Noise Cross-Correlation Matrix:

$$C^{n,n'}_{noise} \equiv \left\langle P_{noise}(\vec{x}_n) P^*_{noise}(\vec{x}_{n'})\right\rangle \cong \sum_{l=1}^{L}\sum_{j=1}^{J} ns^2_{j,l} area_j P(\vec{x}_n, r_j, z_s, \beta_l) P^*(\vec{x}_{n'}, r_j, z_s, \beta_l)$$

$$\vec{x}_n = col(x_n, y_n, z_n),\ n = 1, N_p$$

$$ns^2_{ij} = \text{noise source intensity per unit area}$$

$$area_j = \Delta\beta\ \Delta r\, r_j$$

$$P(\vec{x}_n, r_j, z_s, \beta_l) = \exp[ik_o(r_j - s_{n,l})]u(r_j, z_{ms}, s_{n,l}, z_n, \beta_l)/\sqrt{(r_j - s_{n,l})}$$

$$u(r_j, z_{ms}, s_{n,l}, z_n, \beta_l) = \text{normalized parabolic equation field for a source at range}$$
$$r_j \text{ on bearing } \beta_l \text{ at receiver } \vec{x}_n,\ s_{n,l} = x_n \sin\beta_l + y_n\cos\beta_l$$

Average Omnidirectional Noise Level, $\mathbf{NL_o}$:

$$\mathrm{NL_o} = 10\ \log_{10}\left[trace(C_{noise})/N_p\right],\ \text{dB re 1}\ \mu\text{Pa}^2/\text{Hz}$$

Directional Noise Level, $\mathbf{NL_d}$:

$$\mathrm{NL_d} = 10\ \log_{10}\left[\vec{q}^H(\alpha,\beta) C_{noise}\vec{q}(\alpha,\beta)\right],$$

$$\vec{q}(\alpha,\beta) = \mathrm{col}\left(\exp[-i\vec{k}(\alpha,\beta)\cdot\vec{x}_n], n = 1, N_p\right)$$

$$\vec{k}(\alpha,\beta) = (\omega/c_o)\mathrm{col}(\cos\alpha\sin\beta, \cos\alpha\cos\beta, \sin\alpha)$$

Bounds

$$\mathrm{NL_d} \le \mathrm{NL_o}$$

$$10\ \log_{10}\left[\vec{q}^H(\alpha,\beta) C_{noise}\vec{q}(\alpha,\beta)/N_p^2\right] \le 10\ \log_{10}\left[trace(C_{noise})/N_p\right]$$

References and Suggested Readings

Bannister, R. W., D. J. Kewley, et al. (1989). Directional Underwater Noise Estimates-The Dunes Model," Weapons Systems Research Laboratory, Technical Note WSRL-TN-34/89 (Salisbury, South Australia).

Beilis, A. and F. D. Tappert (1979). "Coupled mode analysis of multiple rough surface scattering." J. Acoust. Soc. Am. 66: 811–826.

Breeding, J. E., L. A. Pflug, et al. (1996). Research Ambient Noise Directionality (RANDI) 3.1 Physics Description, Naval Research Laboratory, NRL/FR/7176- -95-9628 (Stennis Space Center, MS).

Brekhovskikh, L. M. and Y. P. Lysanov (2003). *Fundamentals of Ocean Acoustics, Third Edition*, Springer, New York, NY, p. 136.

Bucker, H. P. (1976). "Use of calculated sound fields and matched-field detection to locate sound sources in shallow water." J. Acoust. Soc. Am. 59: 368–373.

Carey, W. M. (1995). "Standard definitions for sound levels in the ocean." IEEE J. Ocean. Eng. 20: 109–113.

Carey, W. M., R. B. Evans, et al. (1990). "Deep-ocean noise directionality." IEEE J. Ocean. Eng. 15: 324–334.

Carey, W. M. and R. A. Wagstaff (1986). "Low-frequency noise fields." J. Acoust. Soc. Am. 80: 1523–1526.

Cavanagh, R. C. (1974). Fast Ambient Noise Model I (FANM I), The Acoustic Environmental Support Detachment, Office of Naval Research, Washington, DC.

Cavanagh, R. C. and W. W. Renner (1980). "Vertical directionality and depth dependence of averaged acoustic signal and noise." J. Acoust. Soc. Am. 68: 1467–1474.

Chapman, N. R. and J. W. Cornish (1993). "Wind dependence of deep ocean ambient noise at low frequencies." J. Acoust. Soc. Am. 98: 782–789.

Collins, M. D. (1992). "A self-starter for the parabolic equation method." J. Acoust. Soc. Am. 92: 2069–2074.

Collins, M. D. (1993). "A split-step Padé solution for the parabolic equation method." J. Acoust. Soc. Am. 93: 1736–1742.

Courant, R. and D. Hilbert (1953). *Methods of Mathematical Physics, Vol. I.* Interscience, New York, NY, p. 28.

Cron, B. F. and C. H. Sherman (1962). "Spatial-correlation functions for various noise models." J. Acoust. Soc. Am. 34: 1732–1736.

Dashen, R. and W. Munk (1984). "Three models of global ocean noise." J. Acoust. Soc. Am. 76: 540–554.

Dozier, L. B. and F. D. Tappert (1978). "Statistics of normal mode amplitudes in a random ocean. I. Theory." J. Acoust. Soc. Am. 63: 353–365.

Dyer, I. (1973). "Statistics of distant shipping noise." J. Acoust. Soc. Am. 53: 564–570.

Emery, L., M. Bradley, et al. (2001). Historical temporal shipping (HITS) database, Version 4.0, Planning Systems Incorporated, Technical Report, TRS-301 (Slidell, LA).

Etter, P. C. (2001). "Recent advances in underwater acoustic modeling and simulation." J. Sound Vib. 240: 351–383.

Gotlieb, D. and S. Orszag (1977). Numerical Analysis of Spectral Methods: Theory and Applications, CBMS-NSF Regional Conference Series in Applied Mathematics, Vol. 26, SIAM, Philadelphia, pp. 35–38.

Gray, L. M. and D. S. Greeley (1980). "Source level model for propeller blade rate radiation for the world's merchant fleet." J. Acoust. Soc. Am. 67: 516–522.

Hamilton, E. L. (1980). "Geoacoustic modeling of the sea floor." J. Acoust. Soc. Am. 65: 1313–1340.

Hamson, R. M. (1997). "The modeling of ambient noise due to shipping and wind sources in complex environments." Appl. Acoust. 51: 251–287.

Hamson, R. M. and R. W. Wagstaff (1983). An Ambient–noise Model that Includes Coherent Hydrophone Summation for Sonar System Performance in Shallow Water, Saclant ASW Research Centre Report, SR-70 (La Spezia, Italy).

Harrison, C. H. (1997a). "Noise directionality for surface sources in range-dependent environments." J. Acoust. Soc. Am. 102: 2655–2662.

Harrison, C. H. (1997b). "CANARY: A simple model of ambient noise and coherence." Appl. Acoust. 51: 289–315.

Heany, K. D. (2009). "A normal mode projection technique for array response synthesis in range-dependent environments." J. Acoust. Soc. Am. 126: 1036–1045.

Horn, R. A. and C. R. Johnson (1987). *Matrix Analysis*. Cambridge University Press, New York, NY, pp. 169–182.

Jensen, F. B., W. A. Kuperman, et al. (1994). *Computational Ocean Acoustics*. AIP Press, New York, NY, pp. 11–12.

Kewley, D. J., D. G. Browning and W. M. Carey (1990). "Low-frequency wind-generated ambient noise source levels." J. Acoust. Soc. Am. 88: 1894–1902.

Kuperman, W. A. and F. Ingenito (1980). "Spatial correlation of surface generated noise in a stratified ocean." J. Acoust. Soc. Am. 67: 1988–1996.

Lovett, J. R. (1980). "Geographic variation of low-frequency sound absorption in the Atlantic, Indian and Pacific Oceans." J. Acoust. Soc. Am. 67: 338–340.

Maritime-Administration (1995). Commission on the Merchant Marine and Defense, U.S. Maritime Administration.

NAVO (2004). Naval Oceanographic Office, U.S., http://www.navo.navy.mil, accessed (2004).

Osborne, K. R. (1979). DANES – A Directional Ambient Noise Prediction Model for FLENUMOCEANCEN Ocean Data Systems Technical Report (San Diego, CA).

Perkins, J. S., W. A. Kuperman, et al. (1993). "Modeling ambient noise in three-dimensional ocean environments." J. Acoust. Soc. Am. 93: 739–752.

Pierce, A. D. (1981). *Acoustics: An Introduction to its Physical Principles and Applications*. McGraw Hill, New York, NY, p. 22.

Press, W. H., S. A. Teukolsky, et al. (1992). *Numerical Recipes in Fortran, Second Edition*. Cambridge University Press, New York, NY, pp. 140–146.

Renner, W. W. (1986). Ambient Noise Directionality Estimation System (ANDES) Technical Description, Science Applications Int. Corp, SAIC-86/1645 (McLean, VA).

Ross, D. (1987). *Mechanics of Underwater Noise*. Peninsula Publishing, Los Altos, CA, p. 277.

Ross, D. (2005). "Ship sources of ambient noise." IEEE J. Ocean. Eng. 30: 257–261.

Spofford, C. W. (1979). The ASTRAL Model, Vol. 1: Technical Description, Science Applications Inc., SAI-79-742-WA (McLean, VA).

Stakgold, I. (2000). *Boundary Value Problems of Mathematical Physics, Vol. I*, SIAM, Philadelphia, pp. 107–110.

Talham, R. J. (1964). "Ambient-sea-noise model." J. Acoust. Soc. Am. 36: 1541–1544.

Tappert, F. D. (1977). The parabolic approximation method. In *Wave Propagation and Underwater Acoustics, Lecture Notes in Physics, Vol. 70*. J. B. Keller and J. S. Papadakis, (Eds.), Springer-Verlag, New York, NY, p. 265.

Wagstaff, R. A. (1981). "Low-frequency ambient noise in the deep sound channel-The missing component." J. Acoust. Soc. Am. 69: 1009–1014.

Wagstaff, R. A. (2005). "An ambient noise model for the Northeast Pacific Ocean Basin." IEEE J. Ocean. Eng. 30: 286–294.

Chapter 6
Research Issues and Applications of Oceanic Noise

Summary

Ambient noise calculations have traditionally been of naval interest as necessary sonar performance prediction tools. Performance prediction systems had as their primary purpose the allocation of resources and the estimation of system performance in the development of tactics. Since these prediction system could be site-specific, experimental results obtained by sonar systems could be utilized to provide refinement of the environmental acoustic parameters. Recently ambient noise calculations have had an additional role in the conduct of environmental-acoustic impact assessments.

Maritime nations have a responsibility for the safe, clean utilization of the invaluable ocean resource. Recreational, industrial, fishery, aquaculture, and military uses of the ocean resource require quantitative knowledge of the oceanic noise field and the influence these activities have on the environment and the biological systems. In planning for the future economic utilization of this resource, environmental-acoustic impact assessments will require the estimation of noise levels. The quantitative understanding of the physical, biological, and machine-source mechanisms of radiated sound necessitates the development of acoustic noise models based on physical principles that span multiple disciplines.

Knowledge of natural physical production of sound in the ocean has rapidly increased in the last several decades. Theoretical descriptions of transient radiation from microbubbles, microbubble clouds, spray, splash, rain, and turbulence were discussed in Chapters 3 and 4. These physical mechanisms and mathematical methods presented can be adapted for the purpose of ambient noise calculations; but the numerical representation of these sources as well as the description of the sonic radiation from the air–sea boundary interaction zone are still work in progress. In addition, research is required to develop models of radiated noise from man-made machines.

The propagation of sound from the sources of sound in both deep and shallow waters is a key research issue because it is strongly influenced by the variable oceanography. For example, in most shallow water coastal environments, seasonal cooling and warming along with river outflow, tides, and currents produce a variable sound speed structure in depth, range, and time. Wave-theoretic range-dependent

W.M. Carey, R.B. Evans, *Ocean Ambient Noise*, Underwater Acoustics,
DOI 10.1007/978-1-4419-7832-5_6,

propagation codes such as the wide-angle parabolic equation and Gaussian-beam ray codes have been shown to accurately calculate the sound transmission provided appropriate environmental input parameters are used. Past determination of these parameters used large, cumbersome and inefficient archival databases. Satellite observations, archival information coupled with acoustic, basin bathymetry, oceanographic models could provide a method to determine the spatial and temporal variability of the range-dependent oceanographic characteristics of the waveguide. This type of assimilation could provide the basis for synoptic calculations and development of spatial sampling and sensitivity criteria. Furthermore, parallel, "cloud-like," processing should provide for rapidity. In addition, this type of technology can be adapted to utilize the noise field measured on a horizontal, vertical, or volumetric array to perform an estimation of the modal structure and bottom properties using a variety of inversion algorithms [see Buckingham (1987), Siderius et al. (2006), and Harrison et al. (2009)].

At the higher frequencies (above 1 kHz), the Fast Ambient Noise Model and Ambient Noise Directionality Estimation System used geometric acoustics to calculate the vertical directivity of noise. A variety of time-ranged averaged noise codes used convolution to rapidly estimate omnidirectional levels and the directional response of sonar systems. The computational speed of these earlier codes was retained by use of a hybrid ray-mode propagation code, the adiabatic approximation, and range-averaged propagation loss in the Ambient Noise Directionality Estimation System. A recently published application by Wagstaff (2005) to the Northeast Pacific Basin provides a significant example of the validity of the approach and the ability to compute the persistent directional field characteristics of a large basin.

At lower frequencies, the parabolic approximation to the acoustic wave equation due to Tappert (1977) and further developed by Collins (1992, 1993) provided an alternative to adiabatic normal modes. The method incorporates vertical mode coupling needed to accurately describe downslope conversion and the associated contribution to the vertical directivity. The importance of this downslope effect was shown computationally using the parabolic approximation. Jensen (1994) provided a complete description of the computational results with a parabolic equation code to demonstrate the downslope conversion process, and a summary of experimental results may be found in Carey and Wagstaff (1986).

A new computationally extensive calculation of the noise field for the Northeast Pacific Basin was presented in Chapter 5. This wave-theoretic estimation preserved the amplitude as well as the phase of the acoustic noise field and determined the spatial coherence characteristic and array response. In Chapter 5, the assimilation of satellite wind speed, measured source level, and a physical acoustic source model were combined to produce the wind-driven source level per square meter of surface area shown in Fig. 5.13. This source level per square meter of surface area can be updated with a wind speed to produce a temporal calculation of the noise field. The assimilation, computational speed, and treatment of spatially distributed sources is unique. The azimuthal array response was obtained on a sector-by-sector basis shown in Fig. 5.15. Fall sound speed profiles were calculated with a oceanographic model, bathymetry was from a digital bathymetric database, and bottoms of a silty clay for the basin (0° critical angle) and a sandy silt (20° critical angle) for the slopes were used. These choices of environmental variables are the major source of uncertainty in this computation and the reflectivity of the slopes in particular is largely unknown and important.

The low-frequency (50-Hz) vertical directionality as a function of bearing and depth for the basin shipping noise is shown in Fig. 5.18. The major uncertainty in this calculation is the distribution of ships and the source level of ships as a broad representative class. In recent years the size and variety of commercial ships has increased dramatically. The propulsion of commercial ships has evolved from single propellers to dual propellers and multiple types of drives. This situation is further complicated owing to the diversity and size of offshore work boats, fishing boats, and passenger boats. The problem one has with the radiated noise levels from surface crafts is the paucity of calibrated measurements and analytical treatments of the frequency-dependent noise mechanisms. Certainly, one would expect an increase in both deepwater and shallow water noise due to man's activities; the question is the rate at which this increase is occurring. Is the proposal of Ross (1976, 2005) accurate or does this need to be modified to include changes in the radiation level and efficiency with the type of vessel?

The combined application of theoretical acoustics, computational acoustics, applied ocean physics and assimilated satellite observations to the rapid computation of the ocean noise field was shown to be the feasible; however the potential utility to provide environment assessments still needs to be demonstrated. The essence of an accurate noise calculation is a physical understanding of noise production mechanisms, the complex range dependent oceanographic environment, the range and azimuth variation of the bathymetry along with its geophysical composition, the importance of the air-sea boundary interaction zone and the dynamic shipping radiation characteristics. In summary the research question is: if given an accurate numerical source representation, an assimilated satellite sea surface observations, bathymetry, oceanographic model and a method to calculate sound transmission, how accurately can the coherent noise field be estimated?

Noise Mechanisms

The previous quantification of the natural physical mechanisms responsible for noise production near the sea surface required the transient treatment of the radiation from microbubbles, microbubble clouds, and impact noise from rain, spray, and splash. Theoretical expressions for the natural physical sources of sound provide a basis for developing numerical representations. However, the subsurface turbulent boundary layer and the presence of microbubbles convected to depth by the turbulence structure requires additional consideration owing to differences associated with the thermal stability of the boundary layers above and below the sea surface. Satellite sea surface temperature, wind speed, and whitecaps should provide a basis for determining the stability and thus by inference the requisite parameters to refine the frequency- and wind-speed-dependent source representations.

The characteristics of the complex air–sea boundary interaction zone presented in earlier chapters depended on the Reynolds number (R_e), the Richardson number (R_i), and the effective roughness determined by the integral of the sea surface spectrum. The computations, Chapter 5, used the source models coupled with estimates of the 10-m wind speed derived from specialized microwave radar that used pulses (13.4 GHz) and two spot beams in a circular rotation sensitive to scattering from the air–sea boundary. This system is NASA's Quick Scatterometer (QuikSCAT) and is widely used in determination of high-resolution global wind speeds required for estimates of atmospheric forcing and air–sea interaction climate studies. The technical issue for deep sea and shallow water noise computations is whether this wind speed estimate is sufficient or does one also need to use sea surface temperature and whitecap observations?

Finally, storm noise is an interesting and largely unexplored area of ambient noise research. In particular, lightening has been shown to produce interesting oceanic sound. In storms, very low frequency sound production due to wave interactions and large plunging breakers is another phenomenon to be investigated. Perhaps the reason for the lack of experimental and theoretical results is the difficulty of performing reasonable measurements. Snyder (2008) observed the noise field with omnidirectional hydrophones in the Gulf of Mexico for a period of 14 months, including four hurricanes. His measurements at higher frequencies were dramatically affected by the storms, whereas the lower-frequency noise actually decreased, most likely owing to the dramatic decrease in shipping. With satellite measurement such as shown Fig. 5.13 and a directional array (Fig. 5.15) or a submerged autonomous vehicle towed array, additional storm noise measurements could be possible.

Hydromechanical Sources of Sound

A major research issue is the radiation from ships, fishing vessels, oil platforms, and wind farms. The quantification of sound radiated from the modern ship should be parameterized by tonnage, draft, length, beam, type, and speed. A ship's sound radiation is known to be proportional to the power, number of propellers, number of blades, and rotation of the shaft [Arveson and Vendittis (2000), Gray and Greeley (1980)]. Since the hull and wake shadow the fore and aft directions, and the noise is primarily generated by propeller cavitation, the result is a dipole radiation pattern. The issue is whether a class of ships may be represented by an analytical model with horizontal and vertical directivity. Analytical models could also be developed for oil platforms and wind farms.

The determination of the appropriate source representations for surface ships, oil platforms, and wind farms should start with the use of measurements to develop realistic but simplified representations. An autonomous vehicle towed array system is capable of performing this type of measurement cost effectively because of its ability to vary parameters such as range bearing and depth. Such a towed-array vehicle system was developed to demonstrate and quantify the performance characteristics of adaptive signal processing algorithms, including dynamics [Carey and Holmes (2009), Holmes and Carey (2006)]. This technology provides for an effective tool for the measurement of coherent signal propagation and the directional noise fields. The stability of the system enabled the towing of an array some 40 m long capable of formation of synthetic aperture estimates of the modal horizontal wavenumber spectrum. Model-based extended Kalman filter techniques were found to enable both narrow band and broadband tracking [Sullivan and Carey (2009)]. This type of system can provide a unique measurement capability for the measurement of shipping noise source levels, downslope reflectivity, and shallow water geoacoustic inversions. In the case of shipping source levels, such a vehicle could proceed at depth while the ship proceeds on a steady course and constant bearing such as in current common practice. The precision vehicle navigation can then be used to construct the acoustic source characteristics from the measured field. Likewise, the radiation from any platform, wind turbine piling, or any other fixed industrial activity may be measured. Nichols (2005) illustrated some characteristics from oil exploration machines; however, this information is out of date and simply of historical interest. In general, there appears to be a paucity of this type of information available.

The theoretical treatment of radiated sound from ships has been reviewed by Arveson and Vendittis (2000) based on the work of Gray and Greeley (1980) and Ross (1974, see Ross (2005), 1976). These studies were for smaller vessels of approximately 25,000 dwt, 11,000 hp, and with single propellers of radial dimension less than 5 m. The average current (2005–2010) commercial ship is approximately 55,000 dwt with single propellers between 5 and 9 m in diameter. Since compact radiators can be represented by a multipole expansion and since the near field resulting from the vibrating propeller surface is intractable, can the determination of the radiating component by far-field measurements provide for the development of an elegant but simplified source model? This was the approach used to develop an analytical model of sound radiation from the bubble volume oscillation as the propeller rotates from high to low pressure. Several classes of noise sources such as ships were discussed by Ross (1976, 2005) and simplified analytical models were adequate to describe the far-field radiation; however, it is not obvious that more complex ships and oil exploration devices can be easily characterized. Can the measurement of the radiated far field provide for the development of an elegant but simplified source model? If the radiated sound from ships, platforms, and wind farms can be described by simplified analytical descriptions, then the calculation of the noise field is possible.

Geoacoustic Uncertainty

A major uncertainty in the computation of basin-scale deep water and coastal-scale shallow water is the range-dependent geophysical and derived geoacoustic characteristics. In deep water the primary uncertainty is the reflectivity of the continental slopes and sea mounts. Previous discussion (Chapter 4) has focused on the effects of the slope frequency-dependent reflectivity on the directional characteristics of the noise field. The conversion of high angles to lower angles for sources over the slope coupled with the frequency-dependent reflectivity was found to favor lower frequencies with slope enhancement, the megaphone effect. Although this effect has been well recognized (Carey and Wagstaff 1986), the actual reflectivities of the slopes are either unknown or not publicly available. Measurements of slope-reflected ambient noise or noise from an individual ship with deep-ocean-moored high-resolution arrays such as a Mills cross array (Urick 1983) or the end fire array of an autonomous vehicle could provide a basis for a model-based inversion to characterize the slope reflectivity. Basin-scale calculations for the Philippine Sea (Evans and Carey 2009) illustrate the uncertainty of this unknown by varying critical angles between 10° and 20° for frequencies of 50 and 200 Hz. The effect on the vertical and horizontal directionality was found to be considerable.

The continental shelves and deep but bottom-limited gulfs and seas also require knowledge of the geophysical and derived geoacoustic properties of the bottom. The bathymetry and general bottom composition to depths below the sediment water interface are known to a reasonable degree and can be obtained from digital archives. The fine-scale and roughness parameters are simply unknown and consequently the frequency-dependent reflectivity is very uncertain. Naval applications rely not only on geoacoustic profiles but also on site-specific empirical factors derived from measurements. This type of approach is not cost-effective on the scale required for noise computations. Claerbout (1968) proposed the use of low-frequency surface-generated noise for geophysical applications, whereas Buckingham and Jones (1987) showed under certain environmental conditions the bottom critical angle could be determined. The shallow water environment is usually characterized by winter isospeed or upward refracting profiles, spring profiles with variable surface sound speeds, summer profiles with downward refractions, and fall profiles with variable mixed-layer depths. The case of downward refraction offers an opportunity to determine bottom properties when coastal eddies and internal wave effects are minimal.

The increase in the understanding of multiparameter inversions has enabled the estimation of bottom properties when a receiving array and a receiver are employed such as in the experiment conducted by Holmes et al. (2006) in an area where the general limits on bottom properties are known. The use of ambient noise as the source of sound for these types of inversions has been demonstrated by the work of Harrison (2002), Siderius (2006), and Harrison (2009). If one were to use this inversion of shallow water inversions in conjunction with an autonomous vehicle towed array, a cost-effective survey method would provide the required information for noise calculations. It must be mentioned that shallow water environments can be broadly characterized by the formation of the sediment layering by processes of deposition, volcanic layering, and glacial activity. Not all shallow water bottoms may be characterized by inversions and in some interesting areas only measurements will suffice.

Correlation Issues

The response of an array of hydrophones in the noise field is known to be determined by the space-time correlation properties of the field. Cron and Sherman (1962) developed expressions for these correlation functions assuming ergodic random noise sources for volumetric (isotropic) generated noise and surface-generated noise (anisotropic) for directional sources. The noise field was composed of multiple frequency-dependent components such as distant shipping-generated and local wind-generated noise. Cox (1973) examined the correlative properties of temporally stationary and spatially homogeneous (ergodic) noise fields. He employed spherical harmonics and their series expansion to describe the cross-spectral density between two sensors and its wavenumber projection. Both formulations were found to agree with experimental measurements and were useful in the optimum spacing of array hydrophones. Ferguson (1987) applied the methods of Cron and Cox to determine the directional response of a horizontal array to the ocean noise field, with favorable comparisons with experimental results.

The noise response of a vertical array in a weakly range dependent shallow water waveguide, an adiabatic channel, was used to estimate surface source levels by Burgess and Kewley (1983). The difference between upward- and downward-directed beams was related to a distribution of random sources at the air–sea interface. This technique was employed by Harrison et al. (2001) to obtain bottom reflectivity. Coherent sources of sound with either a horizontal or a vertical array coupled with synthetic aperture methods have been employed to produce wavenumber spectra and estimates of the propagating modal eigenfunctions and eigenvalues [Holmes (2006)]. These methods can also be refined by the use of model-based recursive processing and sophisticated multiparameter inversions.

The ambient noise research issue is the use of random surface noise sources or the radiated noise from a surface craft to estimate the modal field near a vertical array. Certainly, the excitation of a set of modal eigenfunctions depends on the near-surface source, the receiver depth, and the propagation characteristics of the shallow water channel. In the case of a weakly range dependent waveguide with minimal attenuation, the adiabatic coupling of modes applies and near the array one would expect to find each mode to be the sum of identical modes with random phase. However, the depth variation would simply be that of the modal eigenfunction and the correlation between pairs of hydrophones will reflect the commonality. One could use the approach of Cron and Cox to this problem or one could use the expected modal field and its eigenvalues. The resulting correlation function can then be used to estimate the Green's function kernel.

Snyder (2008) estimated the correlation coefficient of a time series of power estimates from measurements with hydrophones moored at a depth of 2,935 m, a water depth of 3,200 m, and spaced 2.56 and 2.29 km. He used eight 1-Hz frequency bands between 25 and 1,000 Hz. The correlation coefficient increased with frequency and was greater than 0.5 for frequencies greater than 200 Hz, with correlation times on the order of hours. This investigation shows that spatially separated receivers have correlated power spectral estimates consistent with a weakly range dependent waveguide and a stationary random process.

Roux et al. (2005) showed theoretically that Green's function between two different locations can be estimated by the cross-correlation of the noise field observed at each location. This development assumes an ergodic noise field (stationary and homogeneous medium), adiabatic propagation with minimal attenuation, and mode stripping. Sabra et al. (2005c) also showed that the derivative of the cross-correlation function can be related to the coherent deterministic arrival time. Attenuation was approximated by a perturbed sound speed rather than a complex wavenumber. These studies are an interesting application of ambient noise. It is not clear whether this approach will be feasible in a noise field that is a complex mixture of distance and local sources as well as the frequency dependence of the dispersive waveguide.

Computational Issues

The final issue is the applied ocean physics and applied mathematics task of producing accurate numerical models of an ocean basin assimilated with observational sea surface temperature, current, and bathymetry to provide the basis for range-dependent calculations with distributed sources. The Northeast Pacific calculation shows that such a process is feasible. The issue is the determination of a wave-theoretic transmission calculation to cover the 0.1–10-kHz range that utilizes variable range–depth sampling bandwidths for rapid accurate calculation.

The major problem one faces in such a calculation is the use of multiple high-resolution digital databases, the linkage of this information, and the construction of range-dependent information to be used in the calculation of the sound transmission from the source of noise to the receiver. The calculations that have been performed to date are basically expectation-value calculations and these may be adequate for the higher frequencies, regions where natural mechanisms dominate. But, as seen in Chapter 3, high-resolution arrays observe a two-component noise field, natural noise and dynamic shipping. The degree of resolution in space-time to which this type of computation can or should be extended is an unresolved issue.

References and Suggested Readings

Arvelo, J. I. (2008). "Robustness and constraints of ambient noise inversion." J. Acoust. Soc. Am. 123(2): 679–686.

Arveson, P. T. (2000). "Radiated noise characteristics of a modern day cargo ship." J. Acoust. Soc. Am. 107(1): 118–129.

Arveson, P. T. and D. J. Vendittis (2000). "Radiated noise characteristics of a morden day cargo ship." J. Acoust. Soc. Am. 107(1): 118–129.

Baggeroer, A. B., E. C. Scheer, et al. (2005). "Statistics and vertical directionality of low-frequency ambient noise at the North Pacific acoustic laboratory site." J. Acoust. Soc. Am. 117(3): 1643–1665.

Brooks, L. A. and P. R. Gerstoft (2009). "Green's function approximation from cross-correlations of 20–100 Hz noise during a tropical storm." J. Acoust. Soc. Am. 125(2): 723–734.

Buckingham, M. J. (1981). "Spatial coherence of wind generated noise in as shallow water sound channel." J. Acoust. Soc. Am. 70: 1412.

Buckingham, M. J. and S. S. Jones (1987). "A new shallow-ocean technique for determining the critical angle of the seabed from the vertical directionality of the ambient noise in the water column." J. Acoust. Soc. Am. 81: 938–948.

Burgess, A. S. and D. J. Kewley (1983). "Wind-generated surface noise source levels in deep water East of Australia." J. Acoust. Soc. Am. 73(1): 201–210.

Carey, W. M. (1986). "Measurement of down-slope sound propagation from a shallow source to a deep ocean receiver." J. Acoust. Soc. Am. 79(1): 49–59.

Carey, W. M. (1990). "Special issue on archival papers." IEEE J. Ocean. Eng. 15(4).

Carey, W. M. (1998). "The determination of signal coherence length based on signal coherence and gain measurements in deep and shallow water." J. Acoust. Soc. Am. 104(2, pt 1): 831–837.

Carey, W. M. (2005). "Special issue on archival papers." IEEE J. Ocean. Eng. 30(2).

Carey, W. M., J. Doutt, et al. (1995). "Shallow-water sound transmission measurements on the New Jersey continental shelf." IEEE J. Ocean. Eng. 20(4): 321–336.

Carey, W. M., I. B. Gereben, et al. (1987). "Measurement of sound propagation downslope to a bottom-limited sound channel." J. Acoust. Soc. Am. 81(2): 244–257.

Carey, W. M., J. D. Holmes, et al. (2009). "The applicability of a small autonomous vehicle towed array system to ocean acoustic measurements and signal processing." J. Acoust. Soc. Am. POMA 4: 070007.

Carey, W. M., J. F. Lynch, et al. (2006). "Sound transmission and spatial coherence in selected shallow water areas: Measurements and theory." J. Comput. Acoust. 14(2): 265–298.

Carey, W. M. and R. A. Wagstaff (1986). "Low-frequency noise fields." J. Acoust. Soc. Am. 80(5): 1522–1526.

Claerbout, J. F. (1968). "Synthesis of a layered medium from acoustic transmission response." Geophysics 33: 264–269.

Collins, M. D. (1992). "A self-stater for the parabolic equation method." J. Acoust. Soc. Am. 92: 2069–2074.

Collins, M. D. (1993). "A split-step pade' solution for the parabolic equation method." J. Acoust. Soc. Am. 93: 1736–1742.

Cox, H. (1973). "Spatial correlation in arbitrary noise fields with application to ambient sea noise." J. Acoust. Soc. Am. 54(5): 1289–1301.

Cron, B. F., B. C. Hassell, et al. (1965). "Comparison of theoretical and experimental values of spatial correlation." J. Acoust. Soc. Am. 37(3): 523–529.

Cron, B. F. and C. H. Sherman (1962). "Spatial-correlation functions for various noise models." J. Acoust. Soc. Am. 34: 1732–1736.

Dean, G. B. (2000). "Long time-base observations of the surf noise." J. Acoust. Soc. Am. 107(2): 758–770.

Epifanio, C. L., J. R. Potter, et al. (1999). "Imaging in the ocean with ambient noise: The ORB experiments." J. Acoust. Soc. Am. 106: 3211–3225.

Evans, R. B. and W. M. Carey (2009). Basin Scale Computation of Vertical and Horizontal Directivity of Underwater Noise, Due to Shipping and Wind. International Conference on Theoretical and Computational Acoustics, World Scientific, Dresden, Germany.

Ferguson, B. G. and D. V. Wyllie (1987). "Comparison of observed and theoretical responses of a horizontal line array to wind-induced noise in the deep ocean." J. Acoust. Soc. Am. 82(2): 601–605.

Gray, L. M. and D. S. Greeley (1980). "Source level model for propeller blade rate radiation for the world's merchant fleet." J. Acoust. Soc. Am. 67: 516–522.

Hamsom, R. (1997). "The modeling of ambient noise due to shipping and wind sources in complex environments." Appl. Acoust. 51: 251–287.

Harrison, C. H. (2005). "Performance and limitations of spectral factorization for ambient noise sub-bottom profiling." J. Acoust. Soc. Am. 118(5): 2913–2923.

Harrison, C. H., R. Brind, and A. Cowley (2009). "Bottom reflection properties deduced from ambient noise: Simulation and experiment." J. Comp. Acoust. 9: 327–345.

Harrison, C. H., R. Brind, et al. (2001). "Bottom reflection properties deduced from ambient noise: Simulation and experiment." J. Comp. Acoust. 9: 327–345.

Harrison, C. H. and M. Siderius (2008). "Bottom profiling by correlating beam-steered noise sequences." J. Acoust. Soc. Am. 123(3): 1282–1296.

Harrison, C. H. and D. G. Simons (2002). "Geoacoustic inversion of ambient noise." J. Acoust. Soc. Am. 112: 1377–1389.

Holmes, J. D., W. M. Carey, et al. (2006). "Results from an autonomous underwater vehicle towed hydrophone array experiment in Nantucket Sound." J. Acoust. Soc. Am., Ex-press Ltrs. 120(2): EL15–21.

Jensen, F. B. (2004). Results from the Elba HF-2003 experiment. Proceedings of the High-Frequency Ocean Acoustics Conference, AIP, La Jolla, CA.

Jensen, F. B., W. A. Kuperman, et al. (1994). *Computational Ocean Acoustics*. American Institute of Physics, Inc., New York, NY.

Koch, R. A. and D. P. Knobles (2005). "Geoacoustic inversion with ships as sources." J. Acoust. Soc. Am. 117(2): 626–637.

Kuperman, W. A., and F. Ingenito (1980). "Spatial correlation of surface generated noise in a stratified ocean." J. Acoust. Soc. Am. 67: 1988–1996.

Lynch, J. F., S. D. Rajan, et al. (1991). "A comparison of broadband and narrowband modal inversions for bottom geoacoustic properties at a site near Corpus Christi, Texas." J. Acoust. Soc. Am. 89(2): 648–665.

Makris, N. (1997). "Where the "Acoustic Daylight" analogy breaks down (A)." J. Acoust. Soc. Am. 102(5): 3104–3104.

McDonald, M. A., J. A. Hilderbrand, et al. (2006). "Increases in deep ocean ambient noise in the North Pacific west of San Nicolas Island, California." J. Acoust. Soc. Am. 120(2): 711–718.

McDonald, M. A., J. A. Hilderbrand, et al. (2008). "A 50 year comparison of ambient ocean near San Clemente Island: A bathymetrically complex coastal region off Southern California." J. Acoust. Soc. Am. 124(4): 1985–1992.

Means, S. L. and R. M. Heitmeyer (2002). "Surf generated source signatures: A comparison of plunging and spilling breakers." J. Acoust. Soc. Am. 112(2): 481–488.

Nichols, R. H. (2005). "Some notable noises: Monsters and machines" (Originally published in the Office of Naval Research Proceedings of the International Workshop on Low Frequency Propagation and Noise, 1974) IEEE J. Ocean. Eng. 30(2) (Special Issue, Archival Papers): 248–256.

Nystuen, J. A., E. Amitai, et al. (2008). "Spatial averaging of the oceanic rainfall variability using underwater sound: Ionian sea rainfall experiment 2004." J. Acoust. Soc. Am. 123(3): 1952–1962.

Rajan, S., J. Doutt, et al. (1998). "Inversion for the compressional wave speed profile of the bottom from synthetic aperture experiments conducted in the Hudson Canyon Area", July. IEEE J. Ocean. Eng. 23(3): 174–187.

Rajan, S. D., J. F. Lynch, et al. (1987). "Perturbative inversion methods for obtaining bottom geoacoustic parameters in shallow water." J. Acoust. Soc. Am. 82(3): 998–1017.

Rickett, J. and J. Claerbout (1999). "Acoustic Daylight imaging via spectral factorization: Helioseismology and reservoir monitoring." In Proceedings of the 69th Annual International Meeting, Society of Exploration Geophysicists, pp. 1675–1678.

Ross, D. (1976). *Mechanics of Underwater Noise*. Pergamon Press, New York, NY.

Ross, D. (1987). *Mechanics of Underwater Noise*. Peninsula Publishing, Los Altos, CA.

Ross, D. H. (2005). "Ship Sources of Ambient Noise (Originally published in the Office of Naval Research Proceedings of the International Workshop on Low Frequency Propagation and Noise, 1974)." IEEE J. Ocean. Eng. 30(2) (Special Issue, Archival Papers): 257–261.

Roux, P. and W. A. Kuperman (2004). "Extracting coherent wavefronts from acoustic ambient noise in the ocean." J. Acoust. Soc. Am. 116: 1995–2003.

Roux, P., K. G. Sabra, et al. (2004). "Ambient noise cross-correlation in free space: Theoretical approach." J. Acoust. Soc. Am. 117: 79–84.

Roux, P., K. G. Sabra, et al. (2005). "Ambient noise cross-correlation in free space: Theoretical approach." J. Acoust. Soc. Am. 117: 79–84.

Rozenfeld, I. and W. M. Carey (2001). "Modeling and analysis of sound transmission in the strait of Korea." IEEE J. Ocean. Eng. 26(4): 809–820.

Sabra, K. G., P. Gerstoft, et al. (2005a). "Extracting time domain Green's function estimates from ambient seismic noise." Geophys. Res. Lett. 32: 1029.

Sabra, K. G., P. Roux, et al. (2005b). "Arrival-time structure of the time-averaged ambient noise cross-correlation function in an oceanic waveguide." J. Acoust. Soc. Am. 117: 164–174.

Sabra, K. G., P. Roux, et al. (2005c). "Emergence rate of the time-domain Greens function from the ambient noise cross-correlation function." J. Acoust. Soc. Am. 118: 3524–3531.

Schmidt, H. (1999). OASES users guide and reference manual. Department of Ocean Engineering, MIT, Boston, MA.

Schuster, G. T. (2001). Theory of daylight/interferometer imaging: tutorial: Session A32. 63rd Meeting of the European Association of Exploration Geophysicists. Extended Abstracts.

Siderius, M., C. Harrison, et al. (2007). Geoacoustic inversion of ambient noise and applications to sonar processing. Pacific Rim Underwater Acoustics Conference, Vancouver, BC, http://pruac.apl.washington.edu/

Siderius, M., C. H. Harrison, et al. (2006). "A passive fathometer technique for imaging seaben layering using ambient noise." J. Acoust. Soc. Am. 120: 1315–1323.

Snyder, M. A. (2008). Long-Term Ambient noise Statistics in the Gulf of Mexico, Naval Oceanographic Office, Stennis Space Center, MS, TR 322, p. 170.

Sullivan, E. J., W. M. Carey, et al. (2009). "Passive synthetic aperture as an experimental tool." J. Acoust. Soc. Am. POMA 4: 070008.

Tappert, F. D. (1977). The parabolic approximation method. In *Wave Propagation and Underwater Acoustics*. J. B. Keller and J. S. Papadakis (Eds.), Springer-Verlag, New York, NY.

Urick, R. J. (1983). *Principles of Underwater Sound for Engineers*, 3rd edition. Peninsula Publishing, Los Altos, CA.

Wagstaff, R. A. (2005). "An ambient noise model for the Northeast Pacific Basin." IEEE J. Ocean. Eng. 30(2) (Special Issue, Archival Papers): 286–294.

Wapenaar, K., D. Draganov, et al. (2002). "Theory of acoustic daylight revisited." Society of Exploration Geophysicists, SEG, International Exposition and 72nd Meeting 2002. TU Delft digital repository [http://repository.tudelft.nl/oai], Netherlands.

Wilson, D. K., G. V. Frisk, et al. (2003). "Measurement and prediction of ultralow frequency ocean ambient noise off the eastern U.S. coast." J. Acoust. Soc. Am. 113(6): 3117–3113.

Appendix A
Solutions to the Inhomogeneous Wave Equation

Derivation of the Inhomogeneous Wave Equation

This appendix presents a derivation of the inhomogeneous wave equation for a fluid with a source of fluctuating mass, external forces, and fluctuating fluid velocities. The Lagrangian (Ψ_L) wave function is related to the Eulerian (Ψ_E) as follows:

$$d\Psi_L/dt = \partial\Psi_E/\partial t + (\tilde{v} \cdot \tilde{\nabla})\Psi_E. \tag{1}$$

Since the fluctuating fluid velocity is assumed to be small, the second term on the right-hand side is negligible, and the Eulerian approach is used.

The derivation starts with the equations for the conservation of mass and momentum with the neglect of the energy equation:

For the conservation of mass, we have

$$d\rho'/dt + \rho'\,\tilde{\nabla} \cdot \tilde{v} = q \tag{2}$$

or

$$d\rho'/dt + \rho'\,\partial v_i/\partial x_i = \partial\rho'/\partial t + \partial\rho'\,v_i/\partial x_i = q\,. \tag{3}$$

For the conservation of momentum, we have

$$d\rho'\,\tilde{v}/dt + \rho'\,\tilde{v}(\tilde{\nabla} \cdot \tilde{v}) = \rho'\tilde{g} + \tilde{f}_e - \tilde{\nabla}p' \tag{4}$$

and the ith component is

$$d\rho'\,v_i/dt + \rho'\,v_i\,\partial v_j/\partial x_j = \rho' g_i + f_{ei} - \partial p'_{ij}/\partial x_j. \tag{5}$$

Consequently, the final momentum equation is

$$\partial\rho' v_i/\partial t + \partial\rho'\,v_i v_j/\partial x_j = \rho' g_i + f_{ei} - \partial p'_{ij}/\partial x_j. \tag{6}$$

This equation states that the rate of change of momentum is equal to the summation of the applied forces. The gravitational force can be ignored. The term $\tilde{f}_e$ is an

W.M. Carey, R.B. Evans, *Ocean Ambient Noise*, Underwater Acoustics,
DOI 10.1007/978-1-4419-7832-5,

externally applied force that most likely results in a momentum exchange across an interface or the oscillation of a rigid body. The term p'_{ij} is a stress tensor that represents the normal stress due to pressure and the viscous shear stresses. This can be written in the following form for Newtonian fluids:

$$p'_{ij} = +p'\delta_{ij} - \mu\, D_{ij} - \mu_1\, \Theta\, \delta_{ij}. \tag{7}$$

If one takes the relationship between the dynamic viscosity μ and the second coefficient of viscosity μ_1 to be linear, $\mu_1 = -2\mu/3$, then the second-order deformation tensor D_{ij} and the dilation Θ can be combined as

$$p'_{ij} = +p'\delta_{ij} - \mu\,[D_{ij} - (2/3)\,\Theta\,\delta_{ij}]. \tag{8}$$

These deformation and dilation quantities are introduced to relate the physical meaning of each quantity to relationships found in hydrodynamic turbulence texts such as Hinze (1959). Although the viscous effects have been included, they generally will be ignored. To obtain a wave equation, one takes the temporal derivative of the continuity equation and a spatial derivative of the momentum equation, as follows:

$$\begin{aligned} &\partial/\partial t\,[\partial\rho' / \partial t + \partial\,\rho'\, v_i\,/\partial x_i = q] \\ &\Rightarrow \partial^2\rho' / \partial t^2 + \partial^2\rho'\, v_i\,/\partial t\partial\, x_i = \partial q/\partial t\,; \end{aligned} \tag{9}$$

$$\partial/\partial x_i\;[\partial\,\rho' v_i/\partial t + \partial\rho'\, v_i v_j/\partial x_j = f_{ei} - \partial p'_{ij}/\partial x_j] \tag{10}$$

$$\Rightarrow \partial^2\,\rho' v_i/\partial t\,\partial x_i + \partial^2\rho'\, v_i v_j/\partial x_i\partial x_j = \partial f_{ei}/\partial x_i - \partial^2 p'_{ij}/\partial x_i\partial x_j. \tag{11}$$

Since $\partial\rho' g_i/\partial x_i = 0$, subtracting the spatial derivative of the momentum equation from the temporal derivative of the continuity equation yields

$$\partial^2\rho'/\partial t^2 = \partial^2\rho'\, v_i v_j/\partial x_i\partial x_j + \partial^2 p'_{ij}/\partial x_i\partial x_j - \partial f_{ei}/\partial x_i + \partial q/\partial t. \tag{12}$$

The variables in the above equation represent steady flow as well as the fluctuating quantities, since the acoustic assumption has not been made. The quantity

$$c^2\partial^2\rho'/\partial x_i^2 = \partial^2 c^2\rho'\delta_{ij}/\partial x_i\partial x_j \tag{13}$$

is then subtracted from both sides of the above equation to yield

$$\begin{aligned} \partial^2\rho'/\partial t^2 - c^2\partial^2\rho'/\partial x_i^2 = {}& \partial^2\rho' v_i v_j/\partial x_i\partial x_j + \partial^2 p'_{ij}/\partial x_i\partial x_j - \partial^2 c^2\rho'\delta_{ij}/\partial x_i\partial x_j \\ & -\partial f_{ei}/\partial x_i + \partial q/\partial t\,. \end{aligned} \tag{14}$$

The resulting equation has a wave equation on the left-hand side and source terms on the right-hand side.

$$\partial^2\rho'/\partial t^2 - c^2\partial^2\rho'/\partial x_i^2 = \partial q/\partial t - \partial f_{ei}/\partial x_i + \partial^2/\partial x_i\partial x_j[\rho' v_i v_j + p'_{ij} - c^2\rho'\delta_{ij}] \tag{15}$$

The derivation has not employed the usual linear acoustic assumption of large and small ordered quantities. The result is a general equation for the propagation of waves when the bulk fluid motion is small and viscous stresses are important. This equation can be simplified for the case of an incompressible fluid with no sources or external forces:

$$\partial^2 \rho' v_i v_j/\partial x_i \partial x_j + \partial^2 P'_{ij}/\partial x_i \partial x_j = 0. \tag{16}$$

If one further assumes $\tilde{v} = \tilde{U} + \tilde{u}$, a sum of a stream velocity and a smaller turbulence term, the density $\rho' \rightarrow \rho_o$, the pressure p, and irrotational flow, $\partial U_i/\partial x_j = 0$, then one finds

$$\partial^2 \rho_o U_i U_j/\partial x_i \partial x_j + 2\partial^2 \rho_o U_i u_j/\partial x_i \partial x_j + \partial^2 P/\partial x_i^2 = 0. \tag{17}$$

This result describes the relationship between the pressure gradient and turbulent fluid flow, and for the case of one-dimensional flow yields $\mathrm{P} = \rho \mathrm{U}^2$.

Returning to the compressible equation,

$$\partial^2 \rho'/\partial t^2 - c^2 \partial^2 \rho'/\partial x_i^2 = \partial q/\partial t - \partial f_{ei}/\partial x_i + \partial^2/\partial x_i \partial x_j [\rho' v_i v_j + p'_{ij} - c^2 \rho' \delta_{ij}]. \tag{18}$$

If one assumes

$$\begin{aligned} &\tilde{v} = \tilde{U} + \tilde{u},\ \partial U_i/\partial x_i = 0, \\ &\rho' = \rho_o + \rho,\ \ \partial \rho_o/\partial t = \partial \rho_o/\partial x_i = 0, \\ &\textit{and}\ p' = p_o + p\ , \end{aligned} \tag{19}$$

where lowercase letters without subscripts are second-order fluctuating quantities with respect to the ambient quantities, $u < U,\ \rho < \rho_o,\ p < P_o$, then substitution yields

$$\begin{aligned} &[\partial^2 \rho_o/\partial t^2 - c^2 \partial^2 \rho_o/\partial x_i^2] + [\partial^2 \rho/\partial t^2 - c^2 \partial^2 \rho/\partial x_i^2] \\ &\quad = [\partial q/\partial t - \partial f_{ei}/\partial x_i] \\ &\qquad + \partial^2/\partial x_i \partial x_j [\rho_o U_i U_j + 2\rho_o U_i u_j + \rho U_i U_j + 2\rho U_i u_j + \rho u_i u_j] \\ &\qquad + \partial^2/\partial x_i \partial x_j [(P_o - c^2 \rho_o)\delta_{ij} + (p - c^2 \rho)\delta_{ij}]. \end{aligned} \tag{20}$$

Grouping terms according to their relative order, one obtains

$$\begin{aligned} &\partial^2 \rho/\partial t^2 - c^2 \partial^2 \rho/\partial x_i^2 = \partial q/\partial t - \partial f_{ei}/\partial x_i + \partial^2/\partial x_i \partial x_j [T_{ij}]; \\ &\textit{with} \\ &T_{ij} = 2\rho U_i u_j + \rho U_i U_j + \rho u_i u_j + (p - c^2 \rho)\delta_{ij}\,,\ \textit{the Lighthill stress tensor.} \end{aligned} \tag{21}$$

A compressible liquid has a relationship between pressure and density (the equation of state), $\mathrm{p} = c^2 \rho$; the condition of irrotational flow, $\partial U_i/\partial x_j = 0$; and a velocity potential function ψ. The result is a relation between the pressure and velocity potential of $p = \mp \rho_o \, \partial\psi/\partial t$ and a relation between the particle velocity and the gradient of the potential of $\tilde{v} = \pm\tilde{\nabla}\psi$. Thus, one has equations for the acoustic pressure and velocity potential, as follows:

$$(1/c^2)\,\partial^2 p/\partial t^2 \;-\; \partial^2 p/\partial x_i^2 = \partial q/\partial t - \partial f_{ei}/\partial x_i + \partial^2/\partial x_i \partial x_j [T_{ij}]\,;$$

$$\partial^2\psi/\partial x_i^2 - (1/c^2)\,\partial^2\psi/\partial t^2 \;=\; (1/i\omega\rho)[\partial q/\partial t - \partial f_{ei}/\partial x_i + \partial^2/\partial x_i \partial x_j [T_{ij}]]\,;$$

$$\partial^2\psi/\partial x_i^2 - (1/c^2)\,\partial^2\psi/\partial t^2 \;=\; -4\pi\, f(\tilde{x}, t). \tag{22}$$

Summary

The conservation of mass: $\partial\rho'/\partial t + \partial\rho' \, v_i/\partial x_i = q$
The conservation of momentum:

$$\partial\, \rho' v_i/\partial t \;+\; \partial\rho' \, v_i v_j/\partial x_j \;=\; \rho' g_i + f_{ei} \;-\; \partial p'_{ij}/\partial x_j$$

$$p'_{ij} = +p'\delta_{ij} \;-\; \mu\,[\,D_{ij} - (\,2/3)\,\Theta\,\delta_{ij}]$$

Combined equation:

$$\partial^2\rho'/\partial t^2 \;=\; \partial^2\rho' \, v_i v_j/\partial x_i \partial x_j + \partial^2 p'_{ij}/\partial x_i \partial x_j - \partial f_{ei}/\partial x_i \quad + \partial q/\partial t$$

Assume: $u < U, \;\; \rho < \rho_o, \;\; p < P_o$

$$\tilde{v} = \tilde{U} + \tilde{u}, \quad \partial U_i/\partial x_i = 0, \quad \rho' = \rho_o + \rho, \quad \partial\rho_o/\partial t = \partial\rho_o/\partial x_i = 0\,, \quad p' = p_o + p$$

The inhomogeneous wave equation:

$$\partial^2\rho/\partial t^2 - c^2\partial^2\rho/\partial x_i^2 = \;\partial q/\partial t - \partial f_{ei}/\partial x_i + \;\partial^2/\partial x_i \partial x_j [T_{ij}].$$

$$T_{ij} = 2\rho U_i u_j + \;\rho U_i U_j + \rho u_i u_j + (p - c^2\rho)\delta_{ij} - \mu(D_{ij} - (2/3)\Theta\delta_{ij})$$

$$(1/c^2)\,\partial^2 p/\partial t^2 \;-\; \partial^2 p/\partial x_i^2 = \partial q/\partial t - \partial f_{ei}/\partial x_i + \partial^2/\partial x_i \partial x_j [T_{ij}]$$

$$\partial^2\psi/\partial x_i^2 - (1/c^2)\,\partial^2\psi/\partial t^2 \;=\; -4\pi\, f(\tilde{x}, t).$$

$$4\pi\, f(\tilde{x}, t) \;=\; (-1/i\omega\rho)[\partial q/\partial t - \partial f_{ei}/\partial x_i + \partial^2/\partial x_i \partial x_j [T_{ij}]].$$

The Retarded Green's Function Solution

The Green's function equation can be written as

$$[\tilde{\nabla}^2 - (1/c^2)\partial^2/\partial t^2] \cdot G(\tilde{r}, \tilde{r}_o, t, t_o) = -4\pi\, \delta(\tilde{r} - \tilde{r}_o)\, \delta(t - t_o)\,. \tag{23}$$

Green's function $G(\tilde{r}, \tilde{r}_o, t, t_o)$ is composed of two parts, $G = g + \chi$: the first is the solution for the free space, whereas the second is a solution of the bounded space with the boundary conditions. The solution to the time-dependent, unbounded equation can be obtained by the use of the Laplace transform:

$$L[\nabla^2 - a^2\partial^2/\partial t^2]g(\tilde{r}, \tilde{r}_o, t, t_o)] = -4\pi\,\delta(\tilde{r} - \tilde{r}_o)\,L[\delta(t - t_o)].$$

The Lapace Transform of g:

$$L[-a^2\partial^2/\partial t^2]g(\tilde{r}, \tilde{r}_o, t, t_o)] = (-a^2)[-\dot{g}(\tilde{r}, \tilde{r}_o, 0) - sg(\tilde{r}, \tilde{r}_o, 0) + s^2 L[g(\tilde{r}, \tilde{r}_o, t, t_o)]. \tag{24}$$

The delta function:

$$-4\pi\,\delta(\tilde{r} - \tilde{r}_o)\,L[\delta(t - t_o)] = -4\pi\,\delta(\tilde{r} - \tilde{r}_o)\exp(-st_o).$$

The Initial Condition, I.C.:

$$g(\tilde{r}, \tilde{r}_o, t, t_o) = 0 \quad t - t_o \leq 0.$$

Thus, with zero initial conditions, the Laplace-transformed equation can be written as

$$[\nabla^2 - a^2 s^2]g_s(\tilde{r}, \tilde{r}_o, s) = -4\pi\,\delta(\tilde{r} - \tilde{r}_o)\exp(-st_o). \tag{25}$$

The solution to the spatial portion of this equation can be accomplished by use of a three-dimensional Fourier transform in rectangular coordinates with a wavenumber vector $\tilde{p}(p_1, p_2, p_3)$ *and the delta function* $\delta(\tilde{r} - \tilde{r}_o) = \prod_{i=1}^{3} \delta(x_i - x_{io})$:

$$F\{[\nabla^2 - a^2 s^2]g_s(\tilde{r}, \tilde{r}_o, s)\} = -4\pi F\{\delta(\tilde{r} - \tilde{r}_o)\}\exp(-st_o);$$

$$\left[-\sum_i p_i^2 - a^2 s^2\right] g_{P,s}(\tilde{p}, s) = -4\pi\exp(-i\tilde{p}\cdot\tilde{x}_o - st_o). \tag{26}$$

Since

$$\sum_i p_i^2 = p^2 \; with\, p = |\tilde{p}(p_1, p_2, p_3)|\,, \tag{27}$$

then

$$g_{P,s}(\tilde{p}, s) = 4\pi\,\mathrm{c}^2\exp(-i\tilde{p}\cdot\tilde{x}_o - \mathrm{s}t_o)/[s^2 + p^2\mathrm{c}^2]. \tag{28}$$

Consequently, the problem is finding the inverse of these transforms.

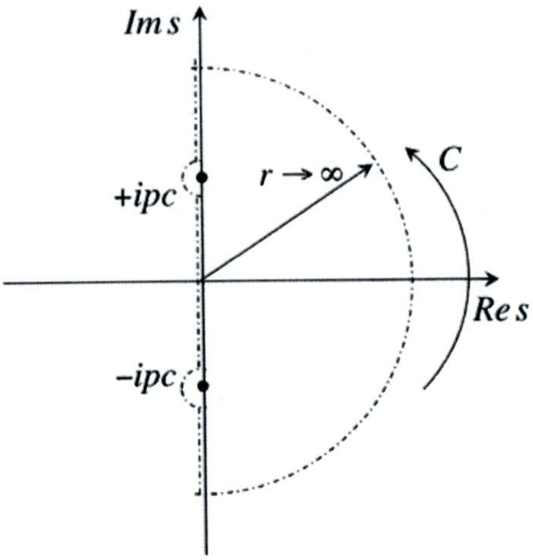

Fig. A.1 Integration contour in the complex s plane

Since

$$g(t) = (1/2\pi i)\int_{\sigma-i\infty}^{\sigma+i\infty} L(g(t))\exp(st)ds = \int_{\sigma-i\infty}^{\sigma+i\infty} g_{p,s}(p,s)\exp(st)ds. \quad (29)$$

Substitution of the expression for $g_{p,s}$ yields

$$\begin{aligned} G_p(p,t) &= (4\pi\, c^2/2\,\pi\, i)\int(\exp(-i\tilde{p}\cdot\tilde{r}_o + s(t-t_o))/(s^2+p^2c^2))ds \\ &= 4\pi\, c^2/2\,\pi\, i)\int(\exp(-i\tilde{p}\cdot\tilde{r}_o + s(t-t_o))/(s+ipc)(s-ipc))ds\,. \end{aligned} \quad (30)$$

This integral may be obtained by integration in the complex s plane using the counters and poles shown in Fig. A.1.
The contour C contains simple poles at$\pm ipc$. Using the theorem of residues, one obtains

$$\begin{aligned} G_p(p,t) &= (4\pi\, c^2/2\,\pi\, i)\int(\exp(-i\tilde{p}\cdot\tilde{r}_o + s(t-t_o))/(s+ipc)(s-ipc))ds \\ &= -2\pi[\exp(-i\tilde{p}\cdot\tilde{r}_o + ipc(t-t_o)) - \exp(-i\tilde{p}\cdot\tilde{r}_o - ipc(t-t_o)]/2ipc \\ &= -2\pi\exp(-i\tilde{p}\cdot\tilde{r}_o)\cdot\sin(pc(t-t_o))/pc\,. \end{aligned} \quad (31)$$

The inverse Fourier transform can be determined:

$$G(\tilde{r}',t') = [-4\pi\, c/(2\pi)^3]\cdot\int\!\!\int\!\!\int\exp(i\tilde{p}\cdot\tilde{r}')\sin(pt'c)/p,$$

where $\tilde{r}' = \tilde{r} - \tilde{r}_o$ *and* $t' = t - t_o$.

$$G(\tilde{r}',t') = [-4\pi\, c/(2\pi)^3]\cdot\int_o^{2\pi}d\phi\int_o^{\pi}\sin(\theta)d\theta\int_o^{\infty}pdp[\exp(i\tilde{p}\cdot\tilde{r}')\sin(pt'c)/p]\,. \quad (32)$$

This integration may be performed by letting $x = \cos(\theta)$ with $-1 \le x \le 1$:

$$G(\tilde{r}', t') = [-8\,c\pi^2/i\,8\,\pi^3]\int_{-1}^{1} dx \int_{o}^{\infty} pdp\ \exp(iprx)\sin(pct')$$
$$= [-c/i\,\pi](1/2ir)\int_{o}^{\infty} dp\,(\exp(iprx) - \exp(iprx))(\exp(ipct') - \exp(ipct'))$$
$$= [c/4\,\pi\,r]\int_{-\infty}^{\infty} dp\,(\exp(iprx) - \exp(iprx))(\exp(ipct') - \exp(ipct'))\,. \tag{33}$$

Substitution of the following definitions,

$$2\,\pi\,\delta(p) \equiv \int_{-\infty}^{\infty} \exp\,(\mathrm{ipx})\mathrm{dx} \text{ and } 2\,\pi\,\delta(\mathrm{x}) \equiv \int_{-\infty}^{\infty} \exp(\mathrm{ipx})\mathrm{dp}, \tag{34}$$

in Eq. (33), yields

$$\mathrm{G}(\tilde{r}', \mathrm{t}') = [\mathrm{c}/2\mathrm{r}] \cdot [\delta(\mathrm{r}' + \mathrm{ct}') + \delta(-\mathrm{r}' - \mathrm{ct}') - \delta(\mathrm{r}' - \mathrm{ct}') - \delta(-\mathrm{r}' + \mathrm{ct}')]. \tag{35}$$

The causality condition $t' = t - t_o \; and \; r' = r - r_o$ eliminates the $\delta(\pm(r' + ct'))$, and since $\delta(r' - ct') = -\delta(-r' + ct')$, one obtains

$$G(r', t') = (c/r')\delta(r' - ct) = (c/\,|r - r_o|)\delta(|r - r_o| - c\,|t - t_o|)\,; \\ = (1/\,|r - r_o|)\delta(|t - t_o| - |r - r_o|\,/c) \tag{36}$$

$|\mathrm{r} - \mathrm{r}_o| = \mathrm{c}\,|\mathrm{t} - \mathrm{t}_o|$ describes the propagation of sound at $\mathrm{r_o}$, and time $\mathrm{t_o}$ to a point r and time t.

Summary

$$[\tilde{\nabla}^2 - (1/\mathrm{c}^2)\partial^2/\partial \mathrm{t}^2] \cdot G(\tilde{\mathrm{r}}, \tilde{\mathrm{r}}_\mathrm{o}, \mathrm{t}, \mathrm{t}_\mathrm{o}) = -4\pi\,\delta(\tilde{\mathrm{r}} - \tilde{\mathrm{r}}_\mathrm{o})\,\delta(\mathrm{t} - \mathrm{t}_\mathrm{o})$$

$$G(\tilde{\mathrm{r}}, \tilde{\mathrm{r}}_\mathrm{o}, \mathrm{t}, \mathrm{t}_\mathrm{o}) = (\mathrm{c}/\,|\mathrm{r} - \mathrm{r}_\mathrm{o}|)\delta\,(|\mathrm{r} - \mathrm{r}_\mathrm{o}| - c\,|\mathrm{t} - \mathrm{t}_\mathrm{o}|)$$

$$= (1/\,|\mathrm{r} - \mathrm{r}_\mathrm{o}|)\delta\,(|\mathrm{t} - \mathrm{t}_\mathrm{o}| - |\mathrm{r} - \mathrm{r}_\mathrm{o}|\,/c)$$

$$G(\tilde{\mathrm{r}}, \tilde{\mathrm{r}}_\mathrm{o}, \mathrm{t}, \mathrm{t}_\mathrm{o}) = \delta\,(\mathrm{t}_\mathrm{o} - \mathrm{t} + |\mathrm{r} - \mathrm{r}_\mathrm{o}|\,/\mathrm{c})/\,|\mathrm{r} - \mathrm{r}_\mathrm{o}|\;\; ; \;\; \mathrm{t} > \mathrm{t}_\mathrm{o}.$$

References and Suggested Readings

Jackson, J. D. (1962). *Classical Electrodynamics*. John Wiley & Sons, Inc., New York, NY, pp. 183–189.

Morse, P. M. and K. U. Ingard (1968). *Theoretical Acoustics*. McGraw-Hill Book Company, New York, NY.

Stratton, J. D. (1941). *Electromagnetic Theory*. McGraw-Hill Book Company, Inc., New York, NY, pp. 424–429.

Solution to the Inhomogeneous Wave Equation

$$[\nabla^2 - (1/c^2)\,\partial^2/\partial t^2]\,\psi(\tilde{r},t,) \;=\; -4\pi\, f(\tilde{r},t). \tag{37}$$

$$f(\tilde{r}(x),t) \;=\; (-1/i4\pi\omega\rho)[\partial q/\partial t - \partial f_{ei}/\partial x_i + \partial^2/\partial x_i\partial x_j[T_{ij}]] \tag{38}$$

$$[\tilde{\nabla}^2 - (1/c^2)\partial^2/\partial t^2]\cdot G(\tilde{r},\tilde{r}_o,t,t_o) = -4\pi\;\delta(\tilde{r}-\tilde{r}_o)\,\delta(t-t_o). \tag{39}$$

Multiply the first equation by $G(\tilde{r},\tilde{r}_o,t,t_o)$ and the second by $\psi(\tilde{r},t)$, subtract, interchange the $\tilde{r},\tilde{r}_o,t,t_o$variables, and integrate.

$$G(\tilde{r},\tilde{r}_o,t,t_o)[\nabla_o^2 - (1/c^2)\,\partial^2/\partial t_o^2]\,\psi(\tilde{r}_o,t_o) \;=\; -4\pi\, f(\tilde{r}_o,t_o)\cdot G(\tilde{r},\tilde{r}_o,t,t_o). \tag{40}$$

$$\psi(\tilde{r}_o,t_o)\cdot[\tilde{\nabla}_o^2 - (1/c^2)\partial^2/\partial t_o^2]\cdot G(\tilde{r},\tilde{r}_o,t,t_o) = -4\pi\;\delta(\tilde{r}-\tilde{r}_o)\,\delta(t-t_o)\psi(\tilde{r}_o,t_o). \tag{41}$$

Dropping the variables as arguments for the sake of brevity, one finds

$$\begin{aligned}&[G\,\nabla_o^2\psi - \psi\nabla_o^2 G] - (1/c^2)[G\partial^2\psi/\partial t_o^2 - \psi\,\partial^2 G/\partial t_o^2]\\ &\quad = -4\pi fG + 4\pi\;\delta(\tilde{r}-\tilde{r}_o)\,\delta(t-t_o)\psi.\end{aligned} \tag{42}$$

Perform an integration with respect to the volume surrounding the source distribution and the field point. The volume V_o is bounded by a surface at infinity with an outward-directed normal n_o. The first term on the left-hand side of the above equation becomes

$$\begin{aligned}&\textstyle\int_{-\infty}^{\infty} dt_o \iiint_{V_o} dV_o[G\nabla_o^2\psi \;-\psi\;\nabla_o^2 G]\\ &\quad = \textstyle\int_{-\infty}^{\infty} dt_o \iint_{S_o} dS_o[g\partial\psi/\partial n_o - \;\psi\;\partial g/\partial n_o] \to 0 \; as\; S_o \to S_\infty.\end{aligned} \tag{43}$$

The second term is integrated with respect to the t_o variable and is

$$\begin{aligned}&\textstyle\iiint_{V_o} dV_o \cdot \int_{-\infty}^{\infty} dt_o(\partial/\partial t_o)[G\partial\psi\;/\partial t_o - \psi\;\partial G/\partial t_o]\\ &\quad \to \textstyle\iiint_{V_o} dV_o \cdot [g\partial\psi\;/\partial t_o - \psi\;\partial g/\partial t_o]_{-\infty}^{\infty} = 0\,.\end{aligned} \tag{44}$$

This results in an expression for the potential function:

$$\begin{aligned}\psi(\tilde{r},t) &= \textstyle\int_{-\infty}^{\infty} dt_o \iiint_{V_o} dV_o[\delta(\tilde{r}-\tilde{r}_o)\delta(t-t_o)\psi(\tilde{r}_o,t_o)]\\ &= \textstyle\int_{-\infty}^{\infty} dt_o \iiint_{V_o} dV_o[f(\tilde{r}_o,t_o)g(\tilde{r},\tilde{r}_o,t,t_o)]\,.\end{aligned} \tag{45}$$

This expression can be further simplified by use of Green's function for $t > t_0$:

$$g(\tilde{r}, \tilde{r}_0, t, t_0) = \delta(|t - t_0| - |\tilde{r} - \tilde{r}_0| / c) = \delta(t_o - (t - |\tilde{r} - \tilde{r}_0| / c)). \quad (46)$$

The quantity $t_o = t - |\tilde{r} - \tilde{r}_0| / c$ represents the retarded time t_o and is equal to the observation time t minus the time, $|\tilde{r} - \tilde{r}_o| / c$, it takes the sound to travel from r_o to r or a sound observed at r. Substituting the expression for g and integrating with respect to t_o yields the final expression for the potential function:

$$\begin{aligned}
\psi(\tilde{r}, t) &= \int_{-\infty}^{\infty} dt_0 \iiint_{V_0} dV_0 [f(\tilde{r}_0, t_0) g(\tilde{r}, \tilde{r}_0, t, t_0)] \\
\psi(\tilde{r}, t) &= \int_{-\infty}^{\infty} dt_0 \iiint_{V_0} dV_0 [f(\tilde{r}_0, t_0) \delta(t_0 - (t - |\tilde{r} - \tilde{r}_0| / c)) / |\tilde{r} - \tilde{r}_o|] \quad (47) \\
\psi(\tilde{r}, t) &= \iiint_{V_o} dV_0 [f(\tilde{r}_0, (t - |\tilde{r} - \tilde{r}_0| / c)) / |\tilde{r} - \tilde{r}_o|] \ .
\end{aligned}$$

This equation states that the radiated field described by $\psi(\tilde{r}, t)$ at point $(\tilde{r}, t)$ is simply the volume integral over the region containing the sources of sound.

References and Suggested Readings

Dowling, A. P. and J. E. Ffowcs Williams (1983). *Sound and Sources of Sound.* Ellis Horwood Limited, Halsted Press-John Wiley & Sons, NY, pp. 37–62, 146–166.

Hinze, J. O. (1959). *Turbulence.* McGraw-Hill Book Co., New York, NY.

Hunt, F. V. (1955). "Notes on the exact equations governing the propagation of sound in fluids." J. Acoust. Soc. Am. 27(6): 1019–1039.

Morse, P. M. and K. U. Ingard (1961). "Linear acoustic theory." In *Handbuch der Physik, Band XI/1, Akustik I.* E. Flügge (Ed.), Springer-Verlag, Berlin, Germany, pp. 1–128.

Morse, P. M. and K. U. Ingard (1968). *Theoretical Acoustics.* McGraw-Hill Book Company, New York, NY.

Pierce, A. D. (1981). *Acoustics: An Introduction to its Physical Principles and Applications.* McGraw Hill Inc, New York, NY (Available Acoustical Society of America, Woodbury, NY).

Rayleigh, J. W. S. (1945). *Theory of Sound.* Dover Publications, New York, NY.

Ross, D. (1976). *Mechanics of Underwater Noise.* Pergamon Press, New York, NY (Also available from Peninsula Publishing, Los Altos, CA).

Summary

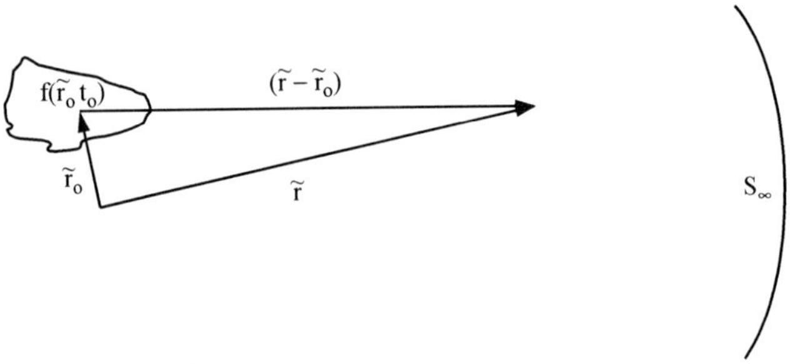

The wave equations:

$$[\nabla^2 - (1/c^2)\,\partial^2/\partial t^2]\,\psi(\tilde{r},t,) \;=\; -4\pi\, f(\tilde{r},t).$$

$$[\tilde{\nabla}^2 - (1/c^2)\partial^2/\partial t^2]\cdot G(\tilde{r},\tilde{r}_o,t,t_o) = -4\pi\,\delta(\tilde{r}-\tilde{r}_o)\,\delta(t-t_o).$$

The integral solution:

$$\begin{aligned}\psi(\tilde{r},t) &= \textstyle\int_{-\infty}^{\infty} dt_o \iiint_{V_o} dV_o[\delta(\tilde{r}-\tilde{r}_o)\delta(t-t_o)\psi(\tilde{r}_o,t_o)]\\ &= (1/4\pi)\textstyle\int_{-\infty}^{\infty} dt_o \iiint_{V_o} dV_o[4\pi)f(\tilde{r}_o,t_o)G\\ &\quad +(G\tilde{\nabla}_o^2\psi(\tilde{r}_o,t_o) - \psi(\tilde{r}_o,t_o)\tilde{\nabla}_o^2 G)\\ &\quad -(\partial/c^2\partial t_o)(G\partial\psi/\partial t_o - \psi\partial G/\partial t_o)]\end{aligned}$$

The case of only a surface at infinity:

$$\int_{-\infty}^{\infty} dt_o \iiint_{V_o} dV_o[G\nabla_o^2\psi \;-\psi\,\nabla_o^2 G] \to 0\,\text{as}\,S_o \to S_\infty.$$

The temporal integral:

$$\iiint_{V_o} dV_o\cdot\int_{-\infty}^{\infty} dt_o(\partial/\partial t_o)[G\partial\psi/\partial t_o - \psi\,\partial G/\partial t_o] = 0.$$

The potential function solution:

$$\begin{aligned}&\psi(\tilde{r},t) = \textstyle\int_{-\infty}^{\infty} dt_o \iiint_{V_o} dV_o[f(\tilde{r}_o,t_o)g(\tilde{r},\tilde{r}_o,t,t_o)]\\ &g(\tilde{r},\tilde{r}_o,t,t_o) = \delta(t_o - (t - |\tilde{r}-\tilde{r}_o|/c))/\,|\tilde{r}-\tilde{r}_o|\\ &\psi(\tilde{r},t) = \textstyle\iiint_{V_o} dV_o[f(\tilde{r}_o,(t-|\tilde{r}-\tilde{r}_o|/c))/\,|\tilde{r}-\tilde{r}_o|]\;.\end{aligned}$$

The Inhomogeneous Wave Equation with a Surface Boundary Condition: Derivation of the Source Integrals

The General Solution

One starts with the inhomogeneous wave equation for pressure and Green's equation with a point source. Using the standard procedure of multiplying one equation, Eq. (48), by Green's function and the other, Eq. (49), by the pressure yields

$$G(\tilde{x},t\mid\tilde{x}_o,t_o)\cdot[\tilde{\nabla}^2-(1/c^2)\,\partial^2/\partial t^2]\cdot P(\tilde{x},t)=-4\pi F(\tilde{x},t)\cdot G(\tilde{x},t\mid\tilde{x}_o,t_o); \quad (48)$$

$$P(\tilde{x},t)\cdot[\tilde{\nabla}^2-(1/c^2)\,\partial^2/\partial t^2]\cdot G(\tilde{x},t\mid\tilde{x}_o,\ t_o)=-4\pi\,\delta(\tilde{x}-\tilde{x}_o)\delta(t-t_o)\cdot P(\tilde{x},t); \quad (49)$$

$$[\circ]=[\tilde{\nabla}^2-(1/c^2)\,\partial^2/\partial t^2]. \quad (50)$$

Subtracting Eq. (49) from Eq. (48) yields

$$G[\circ]P-P[\circ]G=-4\pi F(\tilde{x},t)G(\tilde{x},t\mid\tilde{x}_o,t_o)+4\pi\,\delta(\tilde{x}-\tilde{x}_o)\delta(t-t_o)P(\tilde{x},t). \quad (51)$$

Interchanging $(\tilde{x},t)$ with $(\tilde{x}_o,t_o)$ and integrating with respect to t_o, V_o yields

$$\begin{aligned}P(\tilde{x},t)=&\int_o^{t+\varepsilon}dt_o\iiint_{V_o}dV_oF(\tilde{x}_o,t_o)G(\tilde{x}_o,t_o\mid\tilde{x},t)\\&+(1/4\pi)\int_o^{t+\varepsilon}dt_o\iiint_{V_o}dV_o[G[\circ]P-P[\circ]G]\end{aligned} \quad (52)$$

The first integral on the right-hand side of Eq. (52) may be written as

$$\begin{aligned}\int_o^{t+\varepsilon}dt_o\iiint_{V_o}dV_oF(\tilde{x}_o,t_o)G(\tilde{x}_o,t_o\mid\tilde{x},t)&=\iiint_{V_o}dV_o\int_o^{t+\varepsilon}dt_oF(\tilde{x}_o,t_o)G(\tilde{x}_o,t_o\mid\tilde{x},t)\\&=\iiint_{V_o}(dV_o/r)\cdot[F(\tilde{x}_o,t_o)]_{g,\chi}\end{aligned} \quad (53)$$

where the notation $[\ ]_{g,\chi}$ represents the retardation and the fact that in the presence of the surface $G=g+\chi$. Thus, the presence of a source distribution and the boundary determines the pressure field, $P(\tilde{x},t)$. The remainder of this appendix concerns the reduction of Eq. (52) to account for fundamental acoustic sources and retarded solutions in terms of the initial conditions and boundary conditions.

The Initial Conditions and Surface Boundary Conditions

The second integral on the right-hand side of Eq. (52) contains the surface boundary conditions as well as the initial conditions, and substituting for the operator $[\circ]$ yields

$$\begin{aligned}&(1/4\pi)\int_o^{t+\varepsilon}dt_o\iiint_{V_o}dV_o[G[\circ]P-P[\circ]G]\\&=(1/4\pi)\int_o^{t+\varepsilon}dt_o\iiint_{V_o}dV_o[G(\partial^2/\partial x_{oi}^2)P-P(\partial^2/\partial x_{oi}^2)G]\\&\quad-(1/4\pi c)\int_o^{t+\varepsilon}dt_o\iiint_{V_o}dV_o[G(\partial^2/\partial t_o^2)P-P(\partial^2/\partial t_o^2)G].\end{aligned} \quad (54)$$

The second integral of Eq. (54) may be integrated to yield the statement of initial conditions:

$$
\begin{aligned}
&(1/4\pi c) \int_o^{t+\varepsilon} dt_o \iiint_{V_o} dV_o[G(\partial^2/\partial t_o^2)P - P(\partial^2/\partial t_o^2)G] \\
&\quad = (1/4\pi c) \int_o^{t+\varepsilon} dt_o \iiint_{V_o} dV_o(\partial/\partial t_o)[G(\partial/\partial t_o)P - P(\partial/\partial t_o)G] \\
&\quad = (1/4\pi c) \iiint_{V_o} dV_o[G(\partial/\partial t_o)P - P(\partial/\partial t_o)G]_o^{t+\varepsilon} = 0.
\end{aligned} \tag{55}
$$

The reason this integral is equal to zero is that the upper limit is greater than the interval of the argument of the delta function. The lower limit is the initial condition, which is set to zero because time zero is so far in the distant past that any initial perturbations have died out. The result is the statement

$$
(1/4\pi c) \iiint_{V_o} dV_o[G(\partial/\partial \mathrm{t}_o)P - P(\partial/\partial \mathrm{t}_o)G]_o^{t+\varepsilon} = 0. \tag{56}
$$

The Surface Integrals

The remaining first integral of Eq. (54) is rewritten by use of the divergence theorem to yield a surface integral containing the boundary conditions:

$$
\begin{aligned}
&(1/4\pi) \int_o^{t+\varepsilon} dt_o \iiint_{V_o} dV_o[G(\partial^2/\partial x_{oi}^2)P - P(\partial^2/\partial x_{oi}^2)G] \\
&\quad = (1/4\pi) \int_o^{t+\varepsilon} dt_o \iiint_{V_o} dV_o\partial/\partial x_{oi}[G(\partial/\partial x_{oi})P - P(\partial/\partial x_{oi})G] \\
&\quad = (1/4\pi) \int_o^{t+\varepsilon} dt_o \iint_{S_o} dS_o l_{oi}[G(\partial/\partial x_{oi})P - P(\partial/\partial x_{oi})G].
\end{aligned} \tag{57}
$$

This integral can be simplified further by the following:

$$
\begin{aligned}
&(1/4\pi) \int_o^{t+\varepsilon} dt_o \iint_{S_o} dS_o l_{oi}[G(\partial/\partial x_{oi})P - P(\partial/\partial x_{oi})G] \\
&\quad = (1/4\pi) \iint_{S_o} dS_o l_{oi} \cdot \int_o^{t+\varepsilon} dt_o[G(\partial/\partial x_{oi})P - P(\partial/\partial x_{oi})G] \\
&\quad (1/4\pi) \iint_{S_o} dS_o l_{oi}\{(1/r)\cdot[(\partial/\partial x_{oi})P]_{g,\chi} - \int_o^{t+\varepsilon} dt_o P(\partial/\partial x_{oi})G\}.
\end{aligned} \tag{58}
$$

With G of the form shown in Eq. (36), the integral over time of the spatial derivative can be written as

$$
\begin{aligned}
&\int_o^{t+\varepsilon} dt_o P(\partial/\partial x_{oi})G = \int_o^{t+\varepsilon} dt_o P(\partial/\partial x_{oi})G'/r \\
&\quad = \int_o^{t+\varepsilon} dt_o P[(-1/r^2)(\partial r/\partial x_{oi})G' - (1/r)(\partial r/\partial x_{oi})\partial G'/\partial x_{oi}] \\
&\quad = (-1/r^2)(\partial r/\partial x_{oi})[P]_{g,\chi} - (1/rc)(\partial r/\partial x_{oi})[\partial P/\partial \mathrm{t}_o]_{g,\chi} \; .
\end{aligned} \tag{59}
$$

The final result is

$$
\begin{aligned}
&\rightarrow (1/4\pi) \cdot \iint_{S_o} dS_o\{(l_{oi}/r)[\partial P/\partial x_{oi}]_{g,\chi} \\
&\qquad +(l_{oi}/r^2)(\partial r/\partial x_{oi})[P]_{g,\chi} + (l_{oi}/rc)(\partial r/\partial x_{oi})[\partial P/\partial \mathrm{t}_o]_{g,\chi}\},
\end{aligned} \tag{60}
$$

where $[f]_{g,\chi} = \int_o^{t+\varepsilon} [f\delta(t_o - (t - r_s/c)) + f\delta(t_o - (t - r_i/c))]dt_o$.

This can be simplified by noting that

$$\partial\{[p]/r\}/\partial x_i = (-\partial r/\partial x_i)(p/r^2 + (1/rc)\partial p/\partial t).$$

Upon substitution one has

$$(1/4\pi)\cdot \iint_{S_o} dS_o\{(l_{oi}/r)[\partial P/\partial x_{oi}]_{g,\chi} + l_{oi}\partial\{[p]/r\}/\partial x\},$$

and since $p \to p\delta_{ij}$, the second integral of Eq. (52) is thus reduced to the following:

$$\begin{aligned}&(1/4\pi)\cdot \textstyle\iint_{S_o} dS_o\{(l_{oi}/r)[\partial p\delta_{ij}/\partial x_{oi}]_{g,\chi} + l_{oj}\partial\{[p\delta_{ij}]/r\}/\partial x_i\} =\\ &(1/4\pi)\cdot \textstyle\iint_{S_o} dS_o\{(l_{oi}/r)[\partial p\delta_{ij}/\partial x_{oi}]_{g,\chi} + (1/4\pi)\cdot(\partial/\partial x_i)\iint_{S_o} dS_o l_{oj}[p\delta_{ij}]/r\end{aligned} \tag{61}$$

The Volume Integral Over the Source Region

Returning to the volume integral over the source region, in Eq. (52),

$$\begin{aligned}\textstyle\int_o^{t+\varepsilon} dt_o \iiint_{V_o} dV_o F(\tilde{x}_o, t_o)G(\tilde{x}_o, t_o \mid \tilde{x}, t) &= \textstyle\iiint_{V_o} dV_o \int_o^{t+\varepsilon} dt_o F(\tilde{x}_o, t_o)G(\tilde{x}_o, t_o \mid \tilde{x}, t)\\ &= \textstyle\iiint_{V_o} dV_o/r\cdot[F(\tilde{x}_o, t_o)]_{g,\chi}.\end{aligned} \tag{62}$$

The $F(\tilde{x}_o, t_o)$ term represents the possible sources of hydrodynamic sound and is composed of three general terms:

$$4\pi\, F(\tilde{x}_o, t_o) = \partial q/\partial t_o - \partial f_{ei}/\partial x_{oi} + \partial^2 T_{ij}/\partial x_{oi}\partial x_{oj}. \tag{63}$$

Upon substitution, the volume integral expression becomes

$$\begin{aligned}&\textstyle\iiint_{V_o} dV_o/r\cdot[F(\tilde{x}_o, t_o)]_{g,\chi}\\ &\quad= \textstyle\iiint_{V_o} dV_o/4\pi r\cdot[\partial q/\partial t_o - \partial f_{ei}/\partial x_{oi} + \partial^2 T_{ij}/\partial x_{oi}\partial x_{oj}]_{g,\chi}\\ &\quad= \textstyle\iiint_{V_o} dV_o/4\pi r\cdot[\partial q/\partial t_o]_{g,\chi}\\ &\qquad - \textstyle\iiint_{V_o} dV_o/4\pi r\cdot[\partial f_{ei}/\partial x_{oi}]_{g,\chi}\\ &\qquad + \textstyle\iiint_{V_o} dV_o/4\pi r\cdot[\partial^2 T_{ij}/\partial x_{oi}\partial x_{oj}]_{g,\chi}\ .\end{aligned} \tag{64}$$

(1) The first integral on the right-hand side of Eq. (64), $\iiint_{V_o} dV_o/4\pi r\cdot[\partial q/\partial t_o]_{g,\chi}$, represents the sources of sound associated with a volume pulsation and, consequently, a fluctuating mass such as bubbles and bubble clouds. When compact, these sources are basically monopoles, but because of the sea surface they act as doublets.

(2) The second integral on the right-hand side of Eq. (64), $\iiint_{V_o} dV_o/4\pi r \cdot [\partial f_{ei}/\partial x_{oi}]_{g,\chi}$, can be reduced by use of the following derivative identities:

$$\partial(f_{ei}/r)/\partial x_{oi} = (-1/r^2)(\partial r/\partial x_{oi})f_{ei} + (1/r)(\partial f_{ei}/\partial x_{oi});$$

and

$$\partial(f_{ei}/r)/\partial x_i = (-1/r^2)(\partial r/\partial x_i)f_{ei} + (1/r)(\partial f_{ei}/\partial x_i);$$
$$(\partial r/\partial x_i) = -(\partial r/\partial x_{oi}) \; since \; r = |\tilde{x}_o - \tilde{x}| \; .$$

Because of the difference between the derivative with respect to $\tilde{x}_o$ and $\tilde{x}$, we have

$$\partial f_{ei}(\tilde{x}_o, \tau)/\partial x_{oi} + \partial f_{ei}(\tilde{x}_o, \tau)/\partial x_i = [\partial f_{ei}(\tilde{x}_o, \tau)/\partial x_{oi}]_\tau \; . \tag{65}$$

$$\rightarrow \partial(f_{ei}/r)/\partial x_{oi} + \partial(f_{ei}/r)/\partial x_i = (1/r)[\partial f_e i(\tilde{x}_o, \tau)/\partial x_{oi}]_\tau \; . \tag{66}$$

Consequently, the integral can be written as

$$\begin{aligned} &\iiint_{V_o} dV_o(1/4\pi r)[\partial f_{ei}(\tilde{x}_o, \tau)/\partial x_{oi}]_{g,\chi} \\ &\quad = (1/4\pi) \iiint_{V_o} dV_o\{\partial([f_{ei}]_{g,\chi}/r)/\partial x_{oi} + \partial([f_{ei}]_{g,\chi}/r)/\partial x_i\}. \\ &\quad \rightarrow (1/4\pi) \iint_{S_o} dS_o([f_{ei}]_{g,\chi}/r)l_{oi} \\ &\quad + (1/4\pi)\partial/\partial x_i \iiint_{V_o} ([f_{ei}]_{g,\chi}/r)dV_o. \end{aligned} \tag{67}$$

(3) The third integral on the right-hand side of Eq. (64), $(1/4\pi) \iiint_{V_o} dV_o[\partial^2 T_{ij}/\partial x_{oi}\partial x_{oj}]_{g,\chi}/r$, concerns the stress tensor. In the absence of a surface, this is seen to be a quadrupole source and in the vicinity of the pressure-release surface, it becomes an inefficient higher-order source due to the image. Nevertheless, one can reduce this integral, as follows:

$$(\partial/\partial x_{oi})[(1/r)\partial T_{ij}/\partial x_{oj}] = (-1/r^2)(\partial r/\partial x_{oi})\partial T_{ij}/\partial x_{oj} + (1/r)\partial/\partial x_{oi}(\partial T_{ij}/\partial x_{oj}),$$

and

$$(\partial/\partial x_i)[(1/r)\partial T_{ij}/\partial x_{oj}] = (-1/r^2)(\partial r/\partial x_i)\partial T_{ij}/\partial x_{oj} + (1/r)\partial/\partial x_i(\partial T_{ij}/\partial x_{oj}).$$

Since $\partial r/\partial x_{oi} = -\partial r/\partial x_i$,

$$\begin{aligned} &(\partial/\partial x_{oi})[(1/r)\partial T_{ij}/\partial x_{oj}] + (\partial/\partial x_i)[(1/r)\partial T_{ij}/\partial x_{oj}] = \\ &\quad (1/r)\{\partial/\partial x_{oi}(\partial T_{ij}/\partial x_{oj}) + \partial/\partial x_i(\partial T_{ij}/\partial x_{oj})\} \; . \end{aligned}$$

Because the stress tensor, $T_{ij}(\tilde{x}_o, \tau)$, is only a function of $\tilde{x}_o$, the derivatives in the curly brackets reduce to $(\partial/\partial x_{oi})[\partial T_{ij}/\partial x_{oj}]_\tau$ and by use of the divergence theorem one has

$$(1/4\pi)\iiint_{V_o} dV_o/r\cdot[\partial^2 T_{ij}/\partial x_{oi}\partial x_{oj}]_{g,\chi}$$
$$= (1/4\pi)\iint_{S_o} dS_o(l_{oi}/r)\partial[T_{ij}]/\partial x_{oj}$$
$$+ (1/4\pi)(\partial/\partial x_i)\iiint_{V_o} dV_o(1/r)\,\partial[T_{ij}]/\partial x_{oj}. \tag{68}$$

It remains to reduce the second integral on the left-hand side of Eq. (68). Proceeding as was done previously,

$$\partial[T_{ij}/r]/\partial x_{oj}$$
$$= (-1/r^2)(\partial r/\partial x_{oj})[T_{ij}] + (1/r)[\partial[T_{ij}]/\partial x_{oj} + (\partial[T_{ij}]/\partial\tau)(\partial\tau/\partial r)(\partial r/\partial x_{oj})];$$
$$\partial[T_{ij}/r]/\partial x_j = (-1/r^2)(\partial r/\partial x_j)[T_{ij}] + (1/r)[''0'' + (\partial[T_{ij}]/\partial\tau)(\partial\tau/\partial r)(\partial r/\partial x_j)];$$
$$\partial[T_{ij}/r]/\partial x_{oj} + \partial[T_{ij}/r]/\partial x_j = (1/r)\partial[T_{ij}]/\partial x_{oj}.$$

Substitution in the integral yields the final result:

$$(\partial/\partial x_i)\iiint_{V_o} dV_o(1/r)\,\partial[T_{ij}]/\partial x_{oj}$$
$$= \partial/\partial x_i \iiint dV_o\{\partial[T_{ij}/r]/\partial x_{oj} + \partial[T_{ij}/r]/\partial x_j\} \tag{69}$$
$$= (\partial^2/\partial x_i\partial x_j)\iiint dV_o[T_{ij}]/r + (\partial/\partial x_i)\iint dS_o l_{oj}[T_{ij}]/r\,.$$

The Source Integrals

Collecting the previous results, we obtain the radiated pressure at a position $(\tilde{x}, t)$ as the summation of five source integrals:

$$4\pi\cdot P(\tilde{x}, t) = \sum_{q=1}^{5} I_q(\tilde{x}, t); \tag{70}$$

$$I_1(\tilde{x}, t) = \iiint_{V_o} dV_o/r\cdot[\partial q/\partial \mathrm{t}_o]_{g,\chi}; \tag{71}$$

$$I_2(\tilde{x}, t) = -(\partial/\partial x_i)\iiint_{V_o}([f_{ei}]_{g,\chi}/r)dV_o; \tag{72}$$

$$I_3(\tilde{x}, t) = (\partial^2/\partial x_i\partial x_j)\iiint dV_o[T_{ij}]/r; \tag{73}$$

$$I_4(\tilde{x}, t) = \iint_{S_o} dS_o\{(l_{oi}/r)[\partial p\delta_{ij}/\partial x_{oj}]_{g,\chi} - ([f_{ei}]_{g,\chi}/r)l_{oi} + (l_{oi}/r)\partial[T_{ij}]/\partial x_{oj}\}; \tag{74}$$

$$I_5(\tilde{x}, t) = (\partial/\partial x_i)\iint_{S_o} dS_o\{l_{oj}[p\delta_{ij}]/r + l_{oj}[T_{ij}]/r\}. \tag{75}$$

The final simplification is to use the defining equations

$$f_{ei} = \partial\rho u_i/\partial \mathrm{t}_o + \partial\rho u_i u_j/\partial x_{oj} + \partial p\delta_{ij}/\partial x_{oj}. \tag{76}$$

$$T_{ij} = 2\rho_o U_i u_j + \rho_o u_i u_j + (p - \mathrm{c}^2\rho)\delta_{ij}. \tag{77}$$

$$I_4 \rightarrow \iint dS_o(l_{oi}/r)[\partial \rho u_i/\partial \mathrm{t}_o]_{g,\chi} \quad (78)$$

$$I_5 \rightarrow (\partial/\partial x)_i \cdot \iint dS_o(l_{oi}/r)[2\rho_o U_i u_j + \rho_o u_i u_j + p\delta_{ij}]_{g,\chi}. \quad (79)$$

Summary

(1) The general integral solution:

$$P(\tilde{x}, t) = \int_o^{t+\varepsilon} dt_o \iiint_{V_o} dV_o F(\tilde{x}_o, \mathrm{t}_o) G(\tilde{x}, \mathrm{t}_o | \tilde{x}, t) + (1/4\pi) \int_o^{t+\varepsilon} dt_o \iiint_{V_o} dV_o [G[\mathrm{o}]P - P[\mathrm{o}]G].$$

(2) The volume source integral:

$$(1/4\pi) \iiint_{V_o} dV_o/r \cdot [\partial q/\partial \mathrm{t}_o]_{g,\chi}.$$

(3) The initial condition integral:

$$(1/4\pi c) \iiint_{V_o} dV_o [G(\partial/\partial \mathrm{t}_o]P - P(\partial, \partial \mathrm{t}_o)G]_o^{t+\varepsilon} = 0.$$

(4) The boundary condition integral:

$$(1/4\pi) \int_o^{t+\varepsilon} dt_o \iiint_{V_o} dV_o [G(\partial^2/\partial x_{oi}^2]P - P(\partial^2/\partial x_{oi}^2)G] = (1/4\pi) \cdot \iint_{S_o} dS_o \{(l_{oi}/r)[\partial p\delta_{ij}/\partial x_{oj}]_{g,\chi} + (1/4\pi) \cdot (\partial/\partial x_i) \iint_{S_o} dS_o l_{oj}[p\delta_{ij}]/r\}.$$

(5) The hydrodynamic source function:

$$4\pi\, F(\tilde{x}_o, t_o) = \partial q/\partial \mathrm{t}_o - \partial f_{ei}/\partial x_{oi} + \partial^2 T_{ij}/\partial x_{oi}\partial x_{oj}.$$

(6) The fluctuating external forces:

$$(1/4\pi) \iiint_{V_o} dV_o (1/r)[\partial f_{ei}(\tilde{x}_o, \tau)/\partial x_{oi}]_{g,\chi} = -(1/4\pi) \iint_{S_o} dS_o([f_{ei}]_{g,\chi}/r) l_{oi} + (1/4\pi)\partial/\partial x_i \iiint_{V_o} ([f_{ei}]_{g,\chi}/r) dV_o.$$

(7) The stress tensor integrals:

$$(1/4\pi) \iiint_{V_o} dV_o/r \cdot [\partial^2 T_{ij}/\partial x_{oi}\partial x_{oj}]_{g,\chi}$$
$$= (1/4\pi)(\partial^2/\partial x_i \partial x_j) \iiint dV_o[T_{ij}]/r + (1/4\pi)(\partial/\partial x_i) \iint dS_o l_{oj}[T_{ij}]/r + (1/4\pi) \iint_{S_o} dS_o (l_{oi}/r)\partial[T_{ij}]/\partial x_{oj}.$$

Source Integral Summary

$$4\pi \cdot P(\tilde{x},t) = \sum_{q=1}^{5} I_q(\tilde{x},t)$$

$$I_1(\tilde{x},t) = \iiint_{V_o} dV_o/r \cdot [\partial q/\partial \mathrm{t}_o]_{g,\chi}$$

$$I_2(\tilde{x},t) = (\partial/\partial x_i) \iiint_{V_o} dV_o/r([f_{ei}]_{g,\chi}/r)dV_o$$

$$I_3(\tilde{x},t) = (\partial^2/\partial x_i \partial x_j) \iiint dV_o[T_{ij}]/r$$

$$I_4 = -\iint dS_o(l_{oi}/r)[\partial \rho u_i/\partial t]_{g,\chi}$$

$$I_5 = (\partial/\partial x)_i \cdot \iint dS_o(l_{oi}/r)[2\rho_o U_i u_j + \rho_o u_i u_j + p\delta_{ij}]_{g,\chi}.$$

References and Suggested Readings

Carey, W. M. and D. Browning (1988). "Low frequency ocean ambient noise: Measurements and theory." In *Sea Surface Sound.* B. R. Kerman (Ed.), Kluwer Academic Publishers, Boston, MA, pp. 361–376.

Dowling, A. P. and J. E. Ffowcs Williams (1983). *Sound and Sources of Sound.* Ellis Horwood Limited, Halsted Press-John Wiley & Sons, NY, pp. 37–62, 146–166.

Hinze, J. O. (1959). *Turbulence.* McGraw-Hill Book Co., New York, NY.

Hunt, F. V. (1955). "Notes on the exact equations governing the propagation of sound in fluids." J. Acoust. Soc. Am. 27(6): 1019–1039.

Huon-Li (1981). "On wind-induced underwater ambient noise." NORDA TN 89, NORDA, NSTL, MS.

Jackson, J. D. (1962). *Classical Electrodynamics.* John Wiley & Sons, Inc., New York, NY, pp. 183–189.

Morse, P. M. and K. U. Ingard (1961). "Linear acoustic theory." In *Handbuch der Physik, Band XI/1, Akustik I.* E. Flügge (Ed.), Springer-Verlag, Berlin, Germany, pp. 1–128.

Morse, P. M. and K. U. Ingard (1968). *Theoretical Acoustics.* McGraw-Hill Book Company, New York, NY.

Pierce, A. D. (1981). *Acoustics: An Introduction to its Physical Principles and Applications.* McGraw Hill Inc, New York, NY (Available Acoustical Society of America, Woodbury, NY).

Rayleigh, J. W. S. (1945). *Theory of Sound.* Dover Publications, New York, NY.

Ross, D. (1976). *Mechanics of Underwater Noise.* Pergamon Press, New York, NY (Also available from Peninsula Publishing, Los Altos, CA).

Stratton, J. D. (1941). *Electromagnetic Theory.* McGraw-Hill Book Company, Inc., New York, NY, pp. 424–429.

Appendix B
Standard Definitions

Levels are by definition relative units. In acoustics, the term “level” refers to the logarithm of a nondimensional ratio (R). The logarithm to base 10 ($\log_{10}$) or the natural logarithm (ln) may be used; however, this appendix concerns the use of $\log_{10}$and the decibel.

$$Level \equiv \log(R). \tag{1}$$

In general, the ratio R can be determined for any quantity; however, the *bel* is defined as a unit of level when the logarithmic base is 10 and R is the ratio of two powers (PR):

$$Bel \equiv \log_{10}(PR). \tag{2}$$

It is important to use the ratio of powers or a ratio proportional to the ratio of powers. The decibel (dB) is defined as one tenth of a bel and again requires a ratio of powers:

$$dB \equiv \log_{10^{1/10}}(PR) = 10 \cdot \log_{10}(PR). \tag{3}$$

This definition represents the basic problem acousticians have with the decibel, a level based on a power ratio. Simply put, pressure (μPa) is usually measured and intensity (W/m^2) is usually estimated from the plane wave equation, $I = p_p^2/2\rho c$. When a ratio is formed, $PR = \Pi/\Pi_{ref} = I/I_{ref} = p^2/p_{ref}^2$, the ρc factors cancel and relative level comparisons for sound in the same fluid are valid. However, when the pressure levels in two media, such as air and water, are compared, one must account for this ρc difference. The power ratio for equal pressure amplitudes in air and water is

$$PR = I_{air}/I_{water} = \rho_w c_w/\rho_{air} c_{air} \approx 4.4\,10^3 \rightarrow \approx 36\,dB.$$

Given the power ratio $PR = \Pi\,[W]/\Pi_{ref} = I[W/m^2]/I_{ref} = p^2[\mu Pa]^2/p_{ref}^2$, several reference quantities can be used. The convention used in underwater acoustics is to choose $p_{ref} = 1\,\mu Pa$; this corresponds to $I_{ref} = 0.67 \cdot 10^{-18}\,W/m^2 = 0.67\,aW/m^2$ (a refers to atto). As one may choose any of the above references, the following simple conventions should be used.

$$L_{xxx}\, dB\ re\ units\,(xxx)\ or\ xxx\ Level\, dB\ re\ units\,(xxx)\,. \tag{4}$$

The modifier *xxx* should reflect the reference quantity of power, intensity, or pressure. When the reference quantity is pressure, then one has pressure level dB re 1 μPa and when the reference quantity is intensity, one has intensity level dB re 1 W/m^2.

Harmonic Sound

The fundamental quantity observed or measured in acoustics is the real acoustic pressure. In the case of a simple harmonic *monopole* source, *i.e., a spherical source when the size of the source becomes vanishing small,* the radiated pressure p[μPa], where 1 μPa =10^{-6} N/m^2 to the far field, is an outgoing spherical wave:

$$p(\mathrm{r}, t) = -(ik\rho cS/4\pi)\exp(ik(r-ct))/r = -iQ_s \exp(ik(r-ct))/r \tag{5}$$

Q_s [μPa·m] is the monopole amplitude and ρωS/4π [μ (kg/m^3)(rad/s)(m^3/s)] is the source-strength amplitude. The real pressure is

$$p(\mathrm{r}, t) = (Q_s/r)\sin(k(r-ct) = (Q_s/r)\sin(\omega(r/c-t)). \tag{6}$$

The far-field particle velocity, u(r, t) [m/s], is given by

$$u(\mathrm{r}, t) = (Q_s/\rho cr)\sin(\omega(r/c-t)). \tag{7}$$

The source strength Q_s is related to the intensity, since in the far field the instantaneous intensity I(r, t) [W/m^2] is the product of the pressure and particle velocity

$$\begin{aligned} I(\mathrm{r}, t) &= u(r,t)p(r,t) = (Q_s^2/\rho cr^2)\sin^2(\omega(r/c-t)) \\ &= (Q_s^2/2\rho cr^2)(1-\cos(2\omega(r/c-t)). \end{aligned} \tag{8}$$

The time-averaged intensity *I(r)* at a given radial distance *r* is

$$I(r) = (1/T)\cdot \int_o^T I(\mathrm{r}, t)dt = (Q_s^2/2\rho\, cr^2), [\mathrm{W/m^2}]. \tag{9}$$

The quantity *I(r)* is measured at *r* and extrapolated to r = 1 m by correcting for spreading or transmission loss, to yield

$$I(r = 1m) = (Q_s^2/2\rho\, c),\ \ [W/m^2]. \tag{10}$$

This time-averaged intensity is the average rate of energy flow through a unit area normal to the direction of propagation (W/m^2 = J/m^2 · s = N · m/m^{2}s). Letting $p_p = Q_s$ and $u_p = Q_s/\rho cr$, we have

$$I(r) = < I(\mathrm{r}, t) > = p_p u_p/2 = p_p^2/2\rho c = p_{rms}^2/\rho c = \rho c u_p^2/2 = \rho c u_{rms}^2. \quad (11)$$

The inclusion of the particle velocity terms in Eq. (11) is relevant to the measurement of ambient noise with velocity and pressure-gradient sensors.

The source level of a harmonic source is defined as the ratio of total power radiated by the source to a reference power at a distance of 1 m. For an omnidirectional source, one has

$$\mathrm{SL_o} = 10\log[\,\mathrm{I(1m)/I_{ref}}] = 10\log[\,(\mathrm{p_o}/1\ \mu\mathrm{Pa})^2] dB re 1 \mu\mathrm{Pa}\ @\ 1m. \quad (12)$$

If the source is directional, the intensity is measured on the main response axis I_o(1 m) along with the relative directional pattern $d(\theta,\varphi)$:

$$\begin{aligned} W_s &= \textstyle\iint I_o d(\theta,\varphi)\sin(\theta)^2 d\theta d\phi = I_o \cdot 4\pi \cdot di\ ; \\ PR &= I_o \cdot 4\pi \cdot di/I_{ref} \cdot 4\pi = I_o\, di/I_{ref} = (p_o/p_{ref})^2 di\ ; \\ SL_d &= 10\log[(p_o/p_{ref})^2 di] = SL_o + DI\ . \end{aligned} \quad (13)$$

This calculation of intensity is for a continuous harmonic source. In common measurements performed with linear filters or Fourier transforms, the quantity p(r, ω) is observed and the intensity is

$$I(\omega_o, r) = 1/2\mathrm{R}e(p(r,\omega_o)\cdot p(r,\omega_o)^*/\rho c), \ [W/m^2]. \quad (14)$$

This quantity is related to the mean square pressure by Parseval's theorem since for a harmonic source only a single-frequency line occurs. For this reason, the intensity for a harmonic source is not bandwidth-corrected. The reference intensity is usually taken as that corresponding to either a peak or root mean square pressure amplitude of 1 μPa.

$$I(\omega_o, r)/I_{ref} = [(1/2)p^2]/[(1/2)p_{ref}^2] = p_{rms}^2/p_{ref-rms}^2. \quad (15)$$

Transient Sounds

An important category is that of transient or impulsive sounds. The acoustic energy flux E of a transient or an impulse signal with a duration T observed at a distance r can be extrapolated to an equivalent 1-m distance, to yield

$$E[J/m^2] = (1/\rho\, c)\int_{\mathrm{t_o}}^{\mathrm{t_o}+T} p(r=1,t)^2 dt\,[(\mu Pa)^2(s)/(kg/m^3)\cdot(m/s)] = E_x/\rho c \quad (16)$$

where E_x is the sound exposure. A ratio proportional to power can be obtained with a reference energy flux of a 1-s gated sine wave with a pressure amplitude of 1 μPa:

$$E_{ref} = E_{xref}/\rho\, c = p_{ref}^2 \cdot t_{ref}/\rho c = [\,1\,\mu Pa^2 \cdot s]/\rho c\,; \quad r = 1m. \qquad (17)$$

The reference could also be chosen to be a gated sine wave with a reference energy flux of $E_{ref} = 1\,\text{J/m}^2$ or a total energy flux of $E_{Tref} = 1\,\text{J}$. This would be the preferred method because it lacks ambiguity. However, common practice, since the ρc factors cancel, is to use $(1\mu Pa)^2 \cdot s$ for the reference, and is an important consideration when comparing levels in two different fluids.

Energy Flux Source Level

The energy flux ratio is proportional to a power ratio; thus,

$$E/E_{ref} = E_x/E_{xref} = \left(\int_{t_o}^{t_o+T} p(\mathrm{r},\mathrm{t})^2 dt\right)/p_{ref}^2 t_{ref}; \qquad r = 1m. \qquad (18)$$

This ratio of the energy flux densities can be written in terms of $1\,\text{J/m}^2$ or $(1\,\mu Pa)^2 \cdot (1\,s)$ because the ρc terms cancel. The energy flux source level is

$$EFSL = 10\log_{10}[E/E_{ref}]\,, \; dB\, re\, ((1\mu Pa^2)\cdot(1s))\; @\; 1\, meter. \qquad (19)$$

If the energy flux reference is taken as $E_{ref} = 1\,J/m^2$, then

$$EFSL = 10\log_{10}[E/E_{ref}]\,, \; dB\, re\, (1\,J/m^2)\; @\; 1\, meter. \qquad (20)$$

At any range r, we can form a ratio of either energy or exposures Equation (19) can be formed to obtain the sound exposure ratio and sound exposure level:

$$SEL = 10\cdot\log_{10}[E_x/E_{xref}] = 10\cdot\log_{10}[E_x/p_{ref}^2 t_{ref}],\; dB\, re\, ((1\,\mu Pa^2)(1\,s)). \quad (21)$$

With $E_{ref} = 1\,J/m^2$,

$$SEL = 10\cdot\log_{10}[E_x/E_{xref}]\,,\; dB\, re\, (1\,J/m^2). \qquad (22)$$

Spectral Density of a Transient

The spectral density of a transient may be obtained by using Fourier transform relationships with the following conventions:

$$p(t) = \int_{-\infty}^{\infty} P(\omega)\exp(-i\omega t)d\omega \quad with \quad P(\omega) = (1/2\pi)\int_{-\infty}^{\infty} p(t)\exp(i\omega t)dt,$$
$$p(t) = \int_{-\infty}^{\infty} \{(1/2\pi)\cdot\int_{-\infty}^{\infty} p(t')\exp(i\omega t')dt' \;\exp(-i\omega t)d\omega. \qquad (23)$$

The continuous version of Parseval's theorem is

$$\int_{-\infty}^{+\infty} |p(t)|^2 dt = 2\pi \cdot \int_{-\infty}^{+\infty} P(\omega) \cdot P(\omega)^* d\omega = \int_{-\infty}^{+\infty} P(f) \cdot P(f)^* df = \rho c E = E_x. \tag{24}$$

The quantity $E_{sd}(f) = |P(f)|^2/\rho c$ is the energy flux spectral density (J/m^2 Hz) and is the preferable designation with a reference quantity of 1 J/m^2 Hz; or, simply, as before $|P(f)|^2$, $[(\mu p)^2 \cdot s/Hz]$. This quantity can be used when normalized by a reference energy spectral density $E_{ref,sd} = (1\,\mu\,Pa)^2 \cdot s/Hz$ to yield the energy flux spectral density level:

$$\begin{aligned} \textit{Energy Flux Spectral Density Level} &= 10 \cdot \log(E(1\,Hz\,band)/((1\mu Pa)^2 \cdot s/Hz)) \\ &= EFSDL\,dB\,re\,((1\mu Pa)^2 \cdot s/Hz). \end{aligned} \tag{25}$$

The Impulse

A useful description of a transient is the impulse I_{imp}, which is defined as

$$I_{imp} = \int_o^{T_o} p(\mathrm{r}, \mathrm{t})dt; \quad [\mu\mathrm{Pa} \cdot \mathrm{s}]. \tag{26}$$

The constant T_o is the time of the first sign reversal after the occurrence of the peak pressure. The pressure $p(\mathrm{r}, t)$ is the pressure measured in the far field. This metric may not have sonar significance, but may have a role in the assessment of transient acoustic pressures on marine life. The logarithmic form is seldom used.

Steady Sounds

Sound which is continuous-bounded, nonperiodic, and stationary poses a problem for Fourier analysis. The Fourier integrals are infinite integrals and, consequently, convergence must be considered. Equation (24), Parseval's theorem, states the problem. If the integral of $|p(t)|^2$ converges, then the integral of $|P(f)|^2$ converges. If the pressure variation is random but stationary, $|p(t)|^2$ does not diminish as $t \to \infty$ and the integral will not converge. However, if we restrict the form of p(t) such that

$$\begin{aligned} p(t) = 0, &\quad -\infty < t < -T/2 \\ &\quad T/2 < t < +\infty \quad , \\ p(t) \neq 0, &\quad -T/2 < t < +T/2 \end{aligned} \tag{27}$$

then for large but finite T we have

$$\begin{aligned} &\textstyle\int_{-\infty}^{+\infty} |p(t)|^2 dt \to \int_{-T/2}^{+T/2} |p(t)|^2 dt = T < p(t)^2 >_T = 2\pi \int_{-\infty}^{+\infty} |P(\omega)|^2 d\omega \\ &= \textstyle\int_{-\infty}^{+\infty} |P(f)|^2 df \end{aligned} \tag{28}$$

where $P(f) = \int_{-T/2}^{+T/2} p(t) \exp(i2\pi f\,t)dt$.

These integrals decrease at a sufficient rate as $\omega = 2\pi f \rightarrow \infty$ to ensure convergence for large but not infinite values of T.

$$
\begin{aligned}
&< p(t)^2 >_T = [2\pi/T]\int_{-\infty}^{+\infty} |P(\omega)|^2/d\omega = [1/T]\int_{-\infty}^{+\infty} |P(f)|^2 df \\
&< p(t)^2 >_T/2\pi = \int_{-\infty}^{+\infty} (|P(\omega)|^2/T)d\omega \ \ and < p(t)^2 >_T = \int_{-\infty}^{+\infty} (|P(f)|^2/T)df \\
&(|P(\omega)|^2/T) \ and \ (|P(f)|^2/T) \ are \ spectral \ densities \ per \ unit \ time.
\end{aligned} \tag{29}
$$

For this case on bounded, nonperiodic, and stationary pressure fluctuations, we can define time-varying means and mean square quantities for large T:

$$
< p(t) >_T = (1/T) \int_{-T/2}^{+T/2} p(t)dt \ \ and \ \ < p(t)^2 >_T = \int_{-\infty}^{+\infty} |P(f)|^2 df \tag{30}
$$

The Autocorrelation Function

The question naturally asked is: What happens as a function of time? The autocorrelation function is useful for this purpose:

$$
\begin{aligned}
&\Gamma_p(\tau) \equiv \lim_{T\to\infty} \left[(1/T)\int_{-T/2}^{+T/2} p(t)p(t+\tau)dt\right] \\
&\rightarrow \Gamma_p(0) \equiv \lim_{T\to\infty} \left[(1/T)\int_{-T/2}^{+T/2} p(t)^2 dt\right] = < p(t)^2 >
\end{aligned} \tag{31}
$$

Since the Fourier transform is a more complete description, we have

$$
F\{\Gamma_p(\tau)\} = (1/2\pi)\int_{-\infty}^{+\infty} \Gamma_p(\tau)\exp(-i\omega\tau)d\tau \rightarrow 2\pi P(\omega)P(\omega)^*/T. \tag{32}
$$

Thus, the Fourier transform of the autocorrelation function is $2\pi/T$ times the spectral density $|P(\omega)|^2$ of p(t).

The Power Spectral Density

Continuing with the stationary but random pressure fluctuation, we need to consider sampling. Each time series of p(t) which is observed is one member, a sample member of the family of all possibilities, the ensemble. Let the ensemble be represented by {p(t)} and $p^j(t)$ be the jth sample of the random process {p(t)}. If the variations of the mean, mean square, and autocorrelation of p(t) exhibit significant variation with time, the process is nonstationary, if they exhibit no significant variations with time, the process is weakly stationary, and if all moments of p(t) show no variation with time, the process is strongly stationary. If the moments are the same for any sample of T seconds' duration independent of time, then the process is ergodic.

For $p^j(t)$, which vanishes everywhere outside the interval $t_1 - T/2 < t < t_1 + T/2$ and there is no dependence on t_1, the average power or average energy over the interval is

$$\Pi^j(T) \equiv E^j(T)/T = (1/T) \int_{t_1-T/2}^{t_1+T/2} p^j(t)^2 dt\,. \tag{33}$$

By Parseval's theorem,

$$\Pi^j(T) \equiv E^j(T)/T = \int_{-\infty}^{+\infty} \{\left|P^j(f)\right|^2/T\}df. \tag{34}$$

$$\mathcal{W}^j(f)_T \equiv 2\left|P^j(f)_T\right|^2/T \rightarrow \Pi^j(T) = \int_o^{\infty} \mathcal{W}_p^j(f)_T df\,. \tag{35}$$

Since $\mathcal{W}^j(f)_T$ is an even function of f, we do not need the negative frequencies, and the factor of 2 in the definition above accounts for the change in the integration limits. We then can define a linear average as

$$< \Pi(T)>_{NT} = 1/N \sum_{j=1}^{N} \Pi^j(T) = 1/N \sum_{j=1}^{N} \int_o^{\infty} \mathcal{W}^j(f)_T df\,. \tag{36}$$

If the process is ergodic, one can perform an ensemble average, to obtain

$$< \Pi\,(T)>_e \;=< \Pi^j(T)>_e \;= \int_o^{\infty} < \mathcal{W}^j(f)_T>_e df. \tag{37}$$

For egodic wide-sense stationary processes, one has

$$< \Pi(T)>_{NT} \;=< \Pi(T)>_e. \tag{38}$$

The expected value of the power spectral density is also related to the covariance function K, which follows as a direct consequence of the Wiener–Khintchine theorem:

$$\mathcal{W}(f)_T = 2\,F\{K\} = 2 \int_{-\infty}^{+\infty} K \exp(i\omega t)dt \;=< (2/T)|F(p(T)|^2>_e\,. \tag{39}$$

Equation (39) states that the Fourier transform of the autocorrelation function is equal to 2/T times the spectral densities. Thus, we have

$$\begin{aligned} \mathcal{W}(f)_T &=< (2/T)|F(p(T)|^2>_e \;=< F\{\Gamma(\tau)\}>_e \\ &\rightarrow < \Gamma(\tau)>_e \;= (1/2) \textstyle\int_{-\infty}^{+\infty} \mathcal{W}(f)_T \exp(+i\omega\,\tau)df = K(\tau)\,. \end{aligned} \tag{40}$$

The final measure of the stationary statistical noise is the power spectral density

$$\begin{aligned}\mathcal{W}(f)_T &= < F\{\Gamma(\tau)\}>_e =< F\{K(\tau)\}>_e \\ &= 2\pi < P(\omega)P(\omega)^* > /T =< P(f)P(f)^* > /T\,.\end{aligned} \quad (41)$$

Thus, the power spectral density level is referenced to 1 W/(m^2 Hz) or $(1\mu Pa)^2$/Hz as the natural units of the measurement.

$$\begin{aligned}\textit{Power Spectral Density Level} &\equiv 10 \cdot \log_{10}[\mathcal{W}(f)_T/\mathcal{W}_{ref}(f)_T] \\ &= 10 \cdot \log_{10}[(< p^2(t)>_T/\Delta f)/((\mu\, Pa)^2/Hz)] \ ; \ dB\, re\, (\mu\, Pa)^2/Hz)\,.\end{aligned} \quad (42)$$

Summary

This appendix has stressed the use of te Système International (SI) of metric units and the use of decibel levels to clearly describe sound levels in the ocean. First, the definition of the decibel according to the national standard was used as the logarithm of a ratio proportional to power. It is recommended that a simple convention be used to clarify the use of levels:

$$L_{xxx}\, dB\ re\ units\,(xxx)\ or\ xxx\ Level\, dB\ re\ units\,(xxx)\,.$$

That is, the label and the reference units should match. Second, an additional means of clarity is to list the actual pressure, intensity, power, and energy with the appropriate SI metric unit. The characterization of a sound source depends on whether it is continuous, transient of shot duration, or a longer duration sonar pulse, and also depends on the repetition rate.

A Brief Note on Parseval's Theorem

The use of the term "Parseval's theorem" is widespread and the question to be asked is: What is its relationship to Plancherel's theorem and to Rayleigh's energy theorem? This appendix presents a brief overview of the differences.

If $f(t) = \sum_{n=-\infty}^{n=+\infty} C_n \exp(-i2\pi nt/T)$ is the Fourier series expansion of f(t), then Parseval's theorem can be written as $\frac{1}{T} \int_{-T/2}^{+T/2} |f(t)|^2 dt = \sum_{-\infty}^{+\infty} |C_n|^2\,.$

Converting the Fourier series to the integral transform, one finds if

$$\begin{aligned}p(t) &= (1/2\pi) \int_{-\infty}^{+\infty} \exp(-i\omega t) \cdot P(\omega) d\omega \\ &= (1/2\pi) \int_{-\infty}^{+\infty} \exp(-i\omega t)[\int_{-\infty}^{+\infty} \exp(-i\omega t') \cdot P(t')dt']d\omega \\ &= \int_{-\infty}^{+\infty} \exp(-i2\pi f t)[\int_{-\infty}^{+\infty} \exp(-i2\pi f t') \cdot P(t')dt']df;\end{aligned}$$

then

$$p(t) = \int_{-\infty}^{+\infty} P(\omega) \exp(-i\omega\, t)dt \quad and \quad P(\omega) = \frac{1}{2\pi} \int_{-\infty}^{+\infty} p(t) \exp(i\omega\, t)d\omega \ ;$$

or

$$p(t) = \int_{-\infty}^{+\infty} P(f) \exp(-i2\pi ft)dt \quad and \quad P(f) = \int_{-\infty}^{+\infty} p(t) \exp(i2\pi ft)df \ .$$

The continuous analogue of the discrete form of Parseval's theorem was due to Rayleigh (1889) and later Plancherel (1910) and can be written as

$$\int_{-\infty}^{+\infty} |P(f)|^2 df = \int_{-\infty}^{+\infty} |p(t)|^2 dt = 2\pi \int_{-\infty}^{+\infty} |P(\omega)|^2 d\omega \ with \ \omega = 2\pi f \ .$$

One must be careful with the derivations of these results, especially since the definitions of the delta function must be considered.

For example, if the delta function represents an impulse in time, then

$$\delta(t) = \int_{-\infty}^{+\infty} \exp(-i\omega\, t)dt \ \leftrightarrow \ 1/2\pi = 1/2\pi \int_{-\infty}^{+\infty} \delta(t) \exp(i\omega\, t)d\omega$$

and

$$\delta(t) = \int_{-\infty}^{+\infty} \exp(-i2\pi f\, t)dt \ \leftrightarrow \ 1 = \int_{-\infty}^{+\infty} \delta(t) \exp(i2\pi f\, t)df \ .$$

The Fourier transform of a delta function in time is a "white" or flat-frequency spectrum. To preserve the area for each incremental $d\omega$, one must use a spectral amplitude factor of $1/2\pi$, whereas for each *df*, an amplitude of 1 must be used.

A second example is a delta function in the frequency domain:

$$\exp(-i2\pi f_o t) = \int_{-\infty}^{+\infty} \delta(f - f_o) \exp(-i2\pi ft)dt \ and \ \delta(f - f_o) = \int_{-\infty}^{+\infty} \exp(i2\pi(f - f_o)t)df.$$

This simply states that a sinusoidal function produces a line in the transform spectrum.

An Engineer's Proof of the Energy Theorem

$$p(t) = \int_{-\infty}^{+\infty} P(\omega)\exp(-i2\pi\omega t)d\omega$$

$$p(t)p(t)^* = \int_{-\infty}^{+\infty} P(\omega)\exp(-i2\pi\omega t)d\omega \cdot \int_{-\infty}^{+\infty} P^*(\omega')\exp\big(-i2\pi\omega' t\big)d\omega'$$

$$\int_{-\infty}^{+\infty} p(t)p(t)^* dt = \int_{-\infty}^{+\infty} dt\{\int_{-\infty}^{+\infty} P(\omega)\exp(-i2\pi\omega t)d\omega \cdot \int_{-\infty}^{+\infty} P^*(\omega')\exp\big(-i2\pi\omega' t\big)d\omega'\}$$

$$= \int_{-\infty}^{+\infty} d\omega \cdot \int_{-\infty}^{+\infty} d\omega' P(\omega) \cdot P^*(\omega')\{\int_{-\infty}^{+\infty} \exp\big(-i2\pi(\omega-\omega')t\big)dt\}$$

$$\int_{-\infty}^{+\infty} d\omega \cdot \int_{-\infty}^{+\infty} d\omega' P(\omega)P^*(\omega')\{\int_{-\infty}^{+\infty} \exp\big(-i2\pi(\omega-\omega')t\big)dt\}$$

$$= \int_{-\infty}^{+\infty} d\omega \cdot \int_{-\infty}^{+\infty} d\omega' P(\omega)P^*(\omega')2\pi\,\delta(\omega-\omega')$$

$$\rightarrow \int_{-\infty}^{+\infty} p(t)p(t)^* dt = 2\pi \int_{-\infty}^{+\infty} P(\omega)P^*(\omega)d\omega \ .$$

Summary

$$dB \equiv \log_{10^{1/10}}(PR) = 10 \cdot \log_{10}(PR).$$

Sound Pressure Level

$$\text{SPL} = 10\log[< \text{p}^2 > / < \mu\text{Pa}^2 >],\ \text{dB re}\ \mu(\text{Pa})^2.$$

Energy Flux Source Level

$$EFSL = 10\log_{10}[E/E_{ref}],\ dB\ re\,((1\mu Pa^2)\cdot(1s))\ @\ 1\,meter.$$

$$EFDSL = 10\,Log_{10}[E/E_{ref}],\ dB\ re\,(1\,J/m^2)\ @\ 1\,meter.$$

Energy Flux Spectral Density Level

$$EFSDL = 10\cdot\log(E(1\,Hz\,band)/((1\mu Pa)^2\cdot s/Hz)).$$

Sound Exposure Level

$$SEL = 10\cdot\log_{10}[E_x/E_{xref}],\ dB\ re\,(1\,J/m^2).$$

Power Spectral Density Level

$$\begin{aligned}\text{PSDL} &\equiv 10\cdot\log_{10}[\mathcal{W}(f)_T/\mathcal{W}_{ref}(f)_T],\ dB\ re\,(1\,W/m^2Hz)\\ &= 10\cdot\log_{10}[(< p^2(t) >_{NT}/\Delta f)/ < p^2_{ref} > /\Delta f_{ref}]\ ;\ dB\ re\,(\mu\,Pa)^2/Hz).\end{aligned}$$

Source Levels with the 1-m Convention
Source Level

$$SL = 10\cdot\log(< p^2 > / < p^2_{ref} >),\quad dB\ re\,(\mu Pa)^2\ @\ 1m.$$

Intensity Source Level

$$ISL = 10\cdot\log(I/I_{ref}),\ dB\ re\,(W/m^2).$$

Power Radiated Source Level

$$\Pi_{rad}SL = 10\cdot Log(\Pi/\Pi_{ref}),\ dB\ re\,(1\,W).$$

Selected References

Bendat, J. S. and A. G. Piersol (1966). *Measurement and Analysis of Random Data*. John Wiley & Sons, New York, NY.
Carey, W. M. (1995). "Standard definitions for sound levels in the ocean." IEEE J. Ocean. Eng. 20(2): 109–113.
Hartley, R. V. L. (1924). "The transmission unit." Elec. Comm. 3(1): 34–42.
Horton, C. W. (1968). *Signal Processing of Underwater Acoustic Waves*. U.S. Government Printing Office, Washington, DC (LOCC No.: 74-603409).
Horton, J. W. (1952). "Fundamentals considerations regarding the use of relative magnitudes (Preface by Standards Committee of the I.R.E.)." Proc. I.R.E. 40(4): 440–444.
Horton, J. W. (1954). "The bewildering decibel." Elec. Eng. 73(6): 550–555.
Horton, J. W. (1959). *Fundamentals of Sonar*. U.S. Naval Institute, Annapolis, MD, pp. 40–72.
Marshal, W. J. (1996). "Descriptors of impulsive signal levels commonly used in underwater acoustics." IEEE J. Ocean. Eng. 21(1): 108–110.
Martin, W. H. (1929). "Decibel-the name for the transmission unit." Bell Syst. Tech. J. 8(1): 1–2.
Middleton, D. (1987). *An Introduction to Statistical Communication Theory*. Peninsula Publishing, Los Altos, CA.
Pierce, A. D. (1989). *Acoustics*. Acoustical Society of America, Woodbury, NY, pp. 54–94.
Sparrow, V. W. (1995). "Comments on standard definitions for sound levels in the ocean." *IEEE J. Ocean. Eng.* 20(4): 367–368.

Applicable Standards

ANSI S1.6-1984, (ASA 53-1984), American National Standard Preferred Frequencies and Band Numbers for Acoustical Measurements, American National Standards Institute, New York, NY, 1984.
ANSI/ASME Y10.11-1984, American National Standard Letter Symbols and Abbreviations for Quantities Used in Acoustics, American National Standards Institute, New York, NY, 1984.
ANSI S1.20-1988, (ASA 75-1988), American National Standard Procedures for Calibration of Underwater Electro-Acoustic Transducers, American National Standards Institute, New York, NY, 1988.
ANSI S1.8-1989, (ASA 84-1989), American National Standard Reference Quantities for Acoustical Levels, American National Standards Institute, New York, NY, 1984.
CEI-IEC 27-3, 1989, International Standard Letter: Symbols to be used in Electrical Terminology Part 3: Logarithmic Quantities and Units, International Electrotechnical Commission, Geneva, Switzerland, 1989.
ANSI S1.1-1994, (ASA 111-1994), American National Standard Acoustical Terminology, American National Standards Institute, New York, NY, 1984.
ANSI S1.1-1994, (ASA 111-1994), American National Standard Acoustical Terminology, American National Standards Institute, New York, NY, 1984.
ANSI S1.6-1984, (ASA 53-1984), American National Standard Preferred Frequencies and Band Numbers for Acoustical Measurements, American National Standards Institute, New York, NY, 1984.

Appendix C
A Review of the Sonic Properties of Bubbly Liquids

The Mallock–Wood Approach

Wood (1932) showed that the sonic speed could be calculated for an air bubble/water mixture by use of the mixture density ρ_m and the compressibility K_m. The mixture can be treated as a continuous medium when the bubble diameter d and the spacing between the bubbles D are much less than the wavelength of sound. At low frequencies for a bubbly mixture with gas volume fraction χ, the sonic speed can be calculated from the thermodynamic definition

$$C_m^2 \equiv dP_m/d\rho_m = (\rho_m K_m)^{-1} \tag{1}$$

The total derivative in this equation indicates that the thermodynamic path is important. Two choices immediately occur, evaluate the derivate or evaluate the density–compressibility product. The mixture density and compressibility may be written using the expectation of the spatially varying void fraction χ:

$$\rho_m = (1-\chi)\,\rho_l + \chi\,\rho_g\,; \quad V_m = V_l + V_g\,.$$

$$K_m = \frac{-1}{V_m}\frac{dV_m}{dP} = \frac{-dV_l}{V_l dP}\frac{V_l}{V_m} + \frac{-dV_g}{V_g dP}\frac{V_g}{V_m} = (1-\chi)K_l + \chi K_g. \tag{2}$$

These equations imply that a state of equilibrium prevails, mixture mass is conserved, the pressure is uniform throughout, and there is no slip between the phases. It follows that in the low-frequency region the sonic speed is given by

$$C_m^{-2} = C_{mlf}^{-2} = \left[(1-x)\,\rho_l + x\rho_g\right]\cdot\left[(1-x)\mathrm{K}_l + x\mathrm{K}_g\right]. \tag{3}$$

Consequently,

$$C_m^{-2} = \frac{(1-x)^2}{C_l^2} + \frac{x^2}{C_g^2} + (x)(1-x)\frac{\rho_g^2 C_g^2 + \rho_l^2 C_l^2}{\rho_l \rho_g C_l^2 C_g^2}. \tag{4}$$

This equation can also be obtained by taking the derivative of P_m with respect to ρ_m.

As will be shown later, this expression for the sonic speed poses the question as to whether the gas bubble compressibility is described by an isothermal or an adiabatic process, especially since the sonic speed of air is known to be adiabatic. However, in the case of an air bubble/water mixture, the controlling physical factor is the transfer of heat generated in the bubble compression to the surrounding liquid. If the heat transfer is rapid, then the bubble oscillation is isothermal, and in the absence of surface tension,

$$dV_g/dP = -V_g/P, \quad K_g = 1/P; \tag{5}$$

as compared with the adiabatic condition, ($PV^{\gamma} = Const.$, $\gamma = specific\ heat\ ratio$),

$$dV_g/dP = -V_g/\gamma P, \quad K_g = 1/\gamma P. \tag{6}$$

Isothermal conditions are most likely to prevail for air bubble/water mixtures owing to the large thermal capacity of water.

Examination of Eq. (4) shows that as $x \to 0$, $C_m^{-2} \to C_l^{-2}$, and as $x \to 1$, $C_m^{-2} \to C_g^{-2}$, as one would expect. The striking characteristic of this equation is the sharp reduction in the sonic speed at small volume fractions, i.e., $x = 0.002 \to C_m = 225 m/s$ as shown in Fig. C.1. The equation can be simplified for volume fractions between 0.002 and 0.94 to yield for the mixture

$$C_{mlf}^2 = \frac{\gamma P}{\rho_l x(1-x)} \to \gamma = 1 \to \frac{P}{\rho_l x(1-x)}, \tag{7}$$

with a minimum at $x = 0.5 \to C_m = 20\, m/s$.

The result can be readily extended for a delta function bubble size distribution, all bubbles with the same radius. The resulting well-known equation for a distribution of bubbles with a resonance frequency ω_o and dampening constant δ is

$$\begin{aligned} C_m^{-2} &= \frac{(1-x)^2}{C_l^2} + \frac{x^2}{C_g^2} + \frac{1}{C_{mlf}^2((1-\omega^2/\omega_o^2)-2i\delta\,\omega/\omega_o^2)} \\ &= \frac{(1-x)^2}{C_l^2} + \frac{x^2}{C_g^2} + \frac{\exp(i\phi)}{C_{mlf}^2\left((1-\omega^2/\omega_o^2)^2+(2\delta\,\omega/\omega_o^2)^2\right)^{1/2}} \\ \tan(\phi) &= 2\delta\,\omega/(\omega_o^2-\omega^2) \end{aligned} \tag{8}$$

where C_{mlf} is given by Eq. (7).

These equations have been well known since the 1950s; however, many investigators have rederived these results. Because of the extensive work performed on bubbly liquids, an extensive bibliography can be found at the end of this appendix. The inherent assumptions used in this thermodynamic approach are as follows:

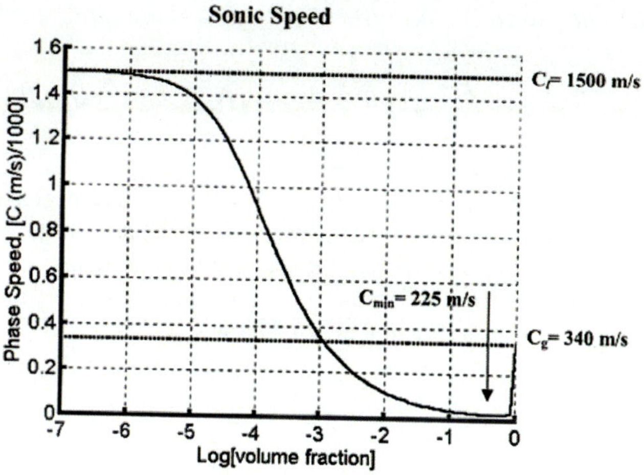

Fig. C.1 The sonic speed versus gas volume fraction for an air bubble/water mixture

Pressure is constant :	$P_m = P_g = P_l = P$
No "slip" :	$U_m = U_b = U_l = U$
Constant mass fraction :	$x\rho_g/(1-x)\rho_l = \text{constant.}$

The neglect of slip, virtual mass, or bubble drag and the evaluation of the dampening factor δ due to thermal, viscous, and dirty effects are subjects of current research.

An alternative derivation and extensive experimental measurements with bubbly liquids were presented by Karplus (1958), Cheng (1985), and Ruggles (1987). These authors started with the expression of the sonic speed in terms of density:

$$C^2 \equiv \frac{dP_m}{d\rho_m}. \tag{9}$$

The density of the mixture is taken as

$$\rho_m = (1-x)\rho_l + x\rho_g. \tag{10}$$

In this equation, it is important to recognize that the volume fraction is the expectation or average value of the volume fraction:

$$x = \iiint \alpha(\tilde{r})dV/V = \iiint V_b(\tilde{r}, a)P(\tilde{r})\,P(a)\,dV/V \tag{11}$$

where $V_b(\tilde{r}, a)$ is the volume of a bubble located at position $\tilde{r}$ with radius a.

This volume is multiplied by the probability of finding a bubble at $\tilde{r}$, $P(\tilde{r})$, and the probability of this bubble having radius a, $P(a)$.

It is important to recognize $dP_m/d\rho_m$ as the total derivative that requires the evaluation along a specific thermodynamic path, the specification of the polytropic process.

$$C_m^{-2} = d\rho_m/dP_m = (1-x)\,d\rho_l/dP_m \; - \rho_l dx/dP_m + \rho_g dx/dP_m + x d\rho_g/dP_m$$
$$\text{with } P_m = P_l = P_b = P$$
$$C_m^{-2} = d\rho_m/dP_m = (1-x)\,d\rho_l/dP \; + \left(\rho_g - \rho_l\right)_g dx/dP + x d\rho_g/dP \tag{12}$$

where $C_l^{-2} = d\rho_l/dP$ *and* $C_g^{-2} = d\rho_g/dP$.

The first assumption, $P_m = P_l = P_b = P$, means we are neglecting surface tension forces as well as any added mass, drag, or dirty bubble effects.

The second assumption here is that the gas bubbles and liquid move with the same velocity. This is termed the "no-slip" assumption with $u_b = u_l = u$. This assumption requires the mass fraction to be constant:

$$\chi\,\rho_g/\,(1-\;\chi)\rho_l = Const. \tag{14 }$$

A consequence of these assumptions is that if the gas obeys the ideal gas law, then the constant-pressure assumption and a constant-temperature or isothermal process yields

$$\frac{\chi\,\rho_g}{(1-\chi)\rho_l} = \frac{\chi\,P_m/RT}{(1-\chi)\rho_l} = Const.\ ;\ and\ \frac{\chi\,P_m}{(1-\chi)\rho_l} = Const. \tag{15}$$

Likewise, when the process is adiabatic,

$$P_g V_g^{\gamma} = P_g(1/\rho_g)^{\gamma} = Const.;\ \ and\ \ \frac{\chi\,P^{1/\gamma}}{(1-\chi)\rho_l} = Const.. \tag{16}$$

In Eq. (12), the quantity needed is $d\chi/dP$.Taking the derivative of Eq. (16), we find

$$d\left[\frac{\chi\,P^{1/\gamma}}{(1-\chi)\rho_l}\right]/dP = 0; \tag{17}$$

$$d\left[\frac{\chi\,P^{1/\gamma}}{(1-\chi)\rho_l}\right]/dP = \frac{\chi}{(1-\chi)}\left[\frac{1}{\rho_l}\frac{dP^{1/\gamma}}{dP} - \frac{P^{1/\gamma}}{\rho_l^2}\frac{d\rho_l}{dP}\right] + \frac{P^{1/\gamma}}{\rho_l}\frac{d[\chi/(1-\chi)]}{dP}. \tag{18}$$

$$d\chi/dP = \frac{\chi}{(1-\chi)}\left[\frac{1}{\rho_l P} - \frac{1}{\rho_l^2}\frac{d\rho_l}{dP}\right] + \frac{d\chi}{dP}\left[\frac{1}{(1-\chi)^2}\right]. \tag{19}$$

$$d\chi/dP = \chi\,(1-\chi)\left[\frac{1}{\rho_l C_l^2} - \frac{1}{\rho_g C_g^2}\right]. \tag{20}$$

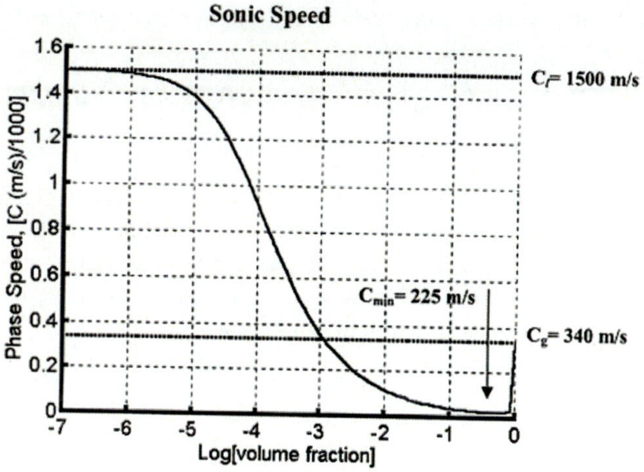

Fig. C.1 The sonic speed versus gas volume fraction for an air bubble/water mixture

Pressure is constant :	$P_m = P_g = P_l = P$
No "slip" :	$U_m = U_b = U_l = U$
Constant mass fraction :	$x\rho_g/(1-x)\rho_l = \text{constant}.$

The neglect of slip, virtual mass, or bubble drag and the evaluation of the dampening factor δ due to thermal, viscous, and dirty effects are subjects of current research.

An alternative derivation and extensive experimental measurements with bubbly liquids were presented by Karplus (1958), Cheng (1985), and Ruggles (1987). These authors started with the expression of the sonic speed in terms of density:

$$C^2 \equiv \frac{dP_m}{d\rho_m}. \tag{9}$$

The density of the mixture is taken as

$$\rho_m = (1-x)\rho_l + x\rho_g. \tag{10}$$

In this equation, it is important to recognize that the volume fraction is the expectation or average value of the volume fraction:

$$x = \iiint \alpha(\tilde{\mathrm{r}})dV/V = \iiint V_b(\tilde{r},a)P\left(\tilde{r}\right)P\left(a\right)dV/V \tag{11}$$

where $V_b(\tilde{r},a)$ is the volume of a bubble located at position $\tilde{r}$ with radius a.

This volume is multiplied by the probability of finding a bubble at $\tilde{r}$, $P(\tilde{r})$, and the probability of this bubble having radius a, $P(a)$.

It is important to recognize $dP_m/d\rho_m$ as the total derivative that requires the evaluation along a specific thermodynamic path, the specification of the polytropic process.

$$C_m^{-2} = d\rho_m/dP_m = (1-x)\,d\rho_l/dP_m \;-\; \rho_l dx/dP_m + \rho_g dx/dP_m + x d\rho_g/dP_m$$
$$\text{with } P_m = P_l = P_b = P$$
$$C_m^{-2} = d\rho_m/dP_m = (1-x)\,d\rho_l/dP \;+\; (\rho_g - \rho_l)_g dx/dP + x d\rho_g/dP \tag{12}$$

where $C_l^{-2} = d\rho_l/dP$ and $C_g^{-2} = d\rho_g/dP$.

The first assumption, $P_m = P_l = P_b = P$, means we are neglecting surface tension forces as well as any added mass, drag, or dirty bubble effects.

The second assumption here is that the gas bubbles and liquid move with the same velocity. This is termed the "no-slip" assumption with $u_b = u_l = u$. This assumption requires the mass fraction to be constant:

$$\chi\,\rho_g/\,(1-\,\chi)\rho_l = Const. \tag{14 }$$

A consequence of these assumptions is that if the gas obeys the ideal gas law, then the constant-pressure assumption and a constant-temperature or isothermal process yields

$$\frac{\chi\,\rho_g}{(1-\chi)\rho_l} = \frac{\chi\,P_m/RT}{(1-\chi)\rho_l} = Const.\ ;\ and\ \frac{\chi\,P_m}{(1-\chi)\rho_l} = Const. \tag{15}$$

Likewise, when the process is adiabatic,

$$P_g V_g^\gamma = P_g(1/\rho_g)^\gamma = Const.;\ \ and\ \frac{\chi\,P^{1/\gamma}}{(1-\chi)\rho_l} = Const.. \tag{16}$$

In Eq. (12), the quantity needed is $d\chi/dP$.Taking the derivative of Eq. (16), we find

$$d\left[\frac{\chi\,P^{1/\gamma}}{(1-\chi)\rho_l}\right]/dP = 0; \tag{17}$$

$$d\left[\frac{\chi\,P^{1/\gamma}}{(1-\chi)\rho_l}\right]/dP = \frac{\chi}{(1-\chi)}\left[\frac{1}{\rho_l}\frac{dP^{1/\gamma}}{dP} - \frac{P^{1/\gamma}}{\rho_l^2}\frac{d\rho_l}{dP}\right] + \frac{P^{1/\gamma}}{\rho_l}\frac{d[\chi/(1-\chi)]}{dP}. \tag{18}$$

$$d\chi/dP = \frac{\chi}{(1-\chi)}\left[\frac{1}{\rho_l P} - \frac{1}{\rho_l^2}\frac{d\rho_l}{dP}\right] + \frac{d\chi}{dP}\left[\frac{1}{(1-\chi)^2}\right]. \tag{19}$$

$$d\chi/dP = \chi\,(1-\chi)\left[\frac{1}{\rho_l C_l^2} - \frac{1}{\rho_g C_g^2}\right]. \tag{20}$$